Modelling Rock Fracturing Processes

Modelling Rock Fracturing Processes

Editor

Anuradha Rawat

scitus
academics

Modelling Rock Fracturing Processes

Edited by **Anuradha Rawat**

Printed in 2017

ISBN: 978-1-68117-148-7

Library of Congress Control Number: 2015936564

SCITUS Academics LLC,
616, Corporate Way, Suite 2, 4766,
Valley Cottage, NY 10989

www.scitusacademics.com

Contents

Preface

This book describes a unique approach using the principles of rock fracture mechanics to investigate the behaviour of fractured rock masses for rock engineering purposes. Rock fracture mechanics, a promising outgrowth of rock mechanics and fracture mechanics, has developed rapidly in recent years, driven by the need for in-depth understanding of rock mass failure processes in both fundamental research and rock engineering designs. Today, as rock engineering extends into many more challenging fields (like mining at depth, radioactive waste disposal, geothermal energy, and deep and large underground spaces), it requires knowledge of rock masses, complex coupled thermal—hydraulic—chemical—mechanical processes. Rock fracture mechanics play a crucial role in these complex coupled processes simply because rock fractures are the principal carrier and common interface. To date, the demand for rock fracture mechanics—based design tools has outstripped the very limited number of numerical tools available.

Editor

Chapter 1

Evaluating the Applicability of Fracture Criteria to Predict the Crack Evolution Path of Dolomite Based on Scb Experiments and Fem

Cunbao Li[1, 2], Lingzhi Xie[1, 2], Li Ren[1, 2], Heping Xie[1, 2], and Jun Wang[1, 2]

[1]Key Laboratory of Energy Engineering Safety and Disaster Mechanics of Ministry of Education, Sichuan University, Chengdu 610065, China

[2]School of Architecture and Environment, Sichuan University, Chengdu 610065, China

ABSTRACT

Mixed mode fracture tests are conducted under various initial loading combinations of mode I and mode II (from pure mode I to pure mode II) on semicircular bend (SCB) specimens of dolomite rock. Damage zones are observed behind the fracture surfaces of the broken samples. Scanning electron microscope images of the fracture surfaces are used to study the failure manner. Using the conventional remesh method based on the finite element method (FEM), several widely accepted fracture criteria are employed to theoretically predict the fracture paths. These criteria include the maximum tangential stress criterion, minimum strain energy density criterion, maximum energy release rate criterion, maximum dilatational strain energy density criterion, and the distortional strain energy density criterion. The applicability of the five fracture criteria is examined. The results show that none of the criteria are successful in predicting the crack trajectories of the predominately mode II cracks; the differences among the predicted results of the crack growth paths are negligible for each crack inclined angle. The effect of Poisson's ratio on the fracture criteria is also investigated and the results show that the predicted crack trajectories are not sensitive to Poisson's ratio.

INTRODUCTION

Extensive development in rock fracture mechanics research has covered many diverse areas including blasting, hydraulic fracturing and in situ stress determination, mechanical fragmentation, rock slope analysis, earthquake mechanics, earthquake prediction, plate tectonics, magmatic intrusions, hot dry rock geothermal energy extraction, fluid transport properties of fracturing rock masses, propagating oceanic rifts, crevasse penetration, and other glaciological problems [1]. As a result of complicated geological processes, numerous flaws, cracks, and faults exist in rock masses. Fracture mechanics is a useful technique for investigating the initiation and propagation of an individual crack or cracks in geological materials. Due to the complexity of in situ stresses, most fractures in these applications occur under a combination of opening and sliding deformation (I/II mixed mode). Under mixed mode loading, a fracture usually propagates along a curvilinear path

[2]. However, it is quite difficult to predict the curvilinear fracture path theoretically because the crack growth direction has to be determined by the present mixed-mode loading condition which corresponds to a specified position on the curvilinear path. Further investigations into fracture initiation angle and fracture propagation path under mixed-mode loading will improve the understanding of mixed mode fracture behavior.

Many researches have been conducted to describe crack initiation and crack propagation under mixed-mode loading. For example, Jia et al. [3] conducted a series of mixed-mode fracture experiments using center cracked circular disc (CCCD) specimens to observe the fracture path of concrete under mixed mode I/II loading; Xeidakis et al. [4] investigated the trajectory of crack growth for marble using antisymmetric three-point bend specimens under different mixed mode loading conditions; Bobet and Einstein [5] studied the fracture initiation, propagation, and, ultimately, coalescence (which plays an important role in the behavior of brittle materials) by loading prefractured specimens of gypsum, used as the rock model material, in uniaxial and biaxial compression; Chen et al. [6] conducted a set of fracture experiments on anisotropic shale by using CCCD specimens under different mixed mode loading conditions for studying the fracture initiation path and also employed the numerical technique of boundary element method to simulate the observed fracture paths. Vásárhelyi and Bobet [7] investigated the crack initiation stress, direction, and propagation of newly generated cracks for both open and closed fractures; Wong et al. [8] numerically investigated the crack initiation, propagation, and coalescence of rock specimens containing preexisting crack-like flaws under compression by using Rock Failure Process Analysis ($RFPA^{2D}$); Al-shayea [9] studied the crack initiation angle and subsequent crack propagation path experimentally for notched Brazilian disk specimens of limestone; Liu et al. [10] used a numerical method for modeling the mixed mode fracture process of heterogeneous rocks using different test samples including CCCD specimens. More recently, Aliha et al. [11] investigated the mixed mode I/II fracture initiation angle and the crack growth trajectory of soft rock (Guiting Limestone) by using both CCCD and edge cracked SCB specimens of various sizes.

In order to theoretically estimate the crack initiation angle, several widely accepted fracture criteria have been proposed, for example, the maximum tangential stress criterion (σ_θ -criterion) [12], minimum strain energy density criterion (S_{min} -criterion) [13], maximum energy release rate criterion (G_{max} -criterion) [14–18], maximum dilatational strain energy density criterion (NT-criterion) [19], and distortional strain energy density criterion (Y-criterion) [20]. From a material mechanics view, it is reasonable to suggest that mixed mode brittle fracture occurs on reaching a certain level of critical energy, tensile stress, tensile strain, shear stress, shear strain, or another physical variable near the crack tip. However, it seems that none of these variables can predict the fracture strengths of predominately mode II fractures accurately because each criterion will predict a corresponding invariant fracture toughness ratio K_{IIc}/K_{Ic} [2] which cannot be applied to a wide range of engineering materials. In fracture mechanics, it is also important to predict where the crack will go, that is, the initiation angle and the growth path of the crack. Therefore, the applicability of these fracture criteria on the crack growth direction and crack evolution path should be discussed.

In this study, a series of mixed-mode fracture tests are conducted on dolomite rock using SCB specimens. Additionally, scanning electron microscope (SEM) images are used to study the failure manner. Moreover, using the conventional remesh method based on the FEM, the crack growth trajectories for various mixed mode loading conditions (from pure mode I to mode II) are theoretically (by the σ_θ -criterion, S_{min} -criterion, G_{max} -criterion, NT-criterion, and Y-criterion) and numerically predicted. A detailed investigation on the applicability of the five fracture criteria to the prediction of crack evolution paths under mixed mode loading conditions has been presented. Finally, the effect of Poisson's ratio on the fracture criteria is also investigated.

CRACK PROPAGATION CRITERIA

Most theoretical fracture criteria consider that when an appropriate characteristic quantity like stress, strain, and energy density energy release rate, in front of the crack tip reaches its critical value which is deemed to a material constant, then the crack begins propagating. For plane problem, the singular stresses around a crack tip in polar coordinate can be expressed as

$$\sigma_{rr} = \frac{1}{\sqrt{2\pi r}}\left[\frac{K_I}{2}\cos\frac{\theta}{2}(3-\cos\theta) - \frac{K_{II}}{2}\sin\frac{\theta}{2}(1-3\cos\theta)\right],$$

$$\sigma_{\theta\theta} = \frac{1}{\sqrt{2\pi r}}\left[\frac{K_I}{2}\cos\frac{\theta}{2}(1+\cos\theta) - \frac{3K_{II}}{2}\sin\frac{\theta}{2}(1+\cos\theta)\right],$$

$$\sigma_{r\theta} = \frac{1}{\sqrt{2\pi r}}\left[\frac{K_I}{2}\sin\frac{\theta}{2}(1+\cos\theta) - \frac{K_{II}}{2}\cos\frac{\theta}{2}(1-3\cos\theta)\right], \tag{1}$$

Where r and θ are the polar coordinates with the origin at the crack tip and K_I and K_{II} are mode I and mode II stress intensity factors, respectively. And the stress field in Cartesian coordinates is defined as

$$\sigma_{xx} = \frac{1}{\sqrt{2\pi r}}\left[K_I\cos\frac{\theta}{2}\left(1-\sin\frac{\theta}{2}\sin\frac{3\theta}{2}\right) - K_{II}\sin\frac{\theta}{2}\left(2+\cos\frac{\theta}{2}\cos\frac{3\theta}{2}\right)\right]$$

$$\sigma_{yy} = \frac{1}{\sqrt{2\pi r}}\left[K_I\cos\frac{\theta}{2}\left(1+\sin\frac{\theta}{2}\sin\frac{3\theta}{2}\right) + K_{II}\sin\frac{\theta}{2}\cos\frac{\theta}{2}\cos\frac{3\theta}{2}\right],$$

$$\sigma_{xy} = \frac{1}{\sqrt{2\pi r}}\left[K_I\cos\frac{\theta}{2}\sin\frac{\theta}{2}\sin\frac{3\theta}{2} + K_{II}\cos\frac{\theta}{2}\left(1-\sin\frac{\theta}{2}\sin\frac{3\theta}{2}\right)\right]$$

$$\sigma_{zz} = 0, \text{ for plane stress,}$$

$$\sigma_{zz} = \nu\left(\sigma_{xx}+\sigma_{yy}\right), \text{ for plane strain,} \tag{2}$$

Where v is Poisson's ratio.

σ_θ-Criterion

Erdogan and Sih [12] propose the maximum tangential stress criterion, which considers that a crack would grow in the direction in which the $\sigma_{\theta\theta}$ is the maximum; the direction should meet the following conditions:

$$\frac{\partial\sigma_{\theta\theta}}{\partial\theta} = 0, \qquad \frac{\partial^2\sigma_{\theta\theta}}{\partial\theta^2} < 0. \tag{3}$$

S_{min}-Criterion

Sih [13] formulates the specific stain energy density factor S at the radius r_0 in term of the energy as follows:

$$S = a_{11}K_I^2 + 2a_{12}K_I K_{II} + a_{22}K_{II}^2, \tag{4}$$

Where

$$a_{11} = \frac{1}{16\pi\mu}(1+\cos\theta)(\kappa-\cos\theta),$$

$$a_{12} = \frac{1}{16\pi\mu}\sin\theta(2\cos\theta-\kappa+1),$$

$$a_{22} = \frac{1}{16\pi\mu}[(\kappa+1)(1-\cos\theta)+(1+\cos\theta)(3\cos\theta-1)], \tag{5}$$

Where μ is the shear modulus, κ=3−4v for the plane strain problem, and κ= (3−4v)/ (1+v) for the plane stress problem. The S_{min} -criterion postulates that the crack initiates in the direction of minimum strain energy density along a constant radius around the crack tip when S reaches a critical value; that is,

$$\frac{\partial S}{\partial \theta} = 0, \qquad \frac{\partial^2 S}{\partial \theta^2} > 0. \tag{6}$$

G_{max} -Criterion

Several forms [15–18] of G_{max} -criterion have been formulated and only the expressions presented by Hussain et al. [15] will be reviewed here. The energy release rate is given as

$$G = \frac{4}{E}\left(\frac{1}{3+\cos^2\theta}\right)^2\left(\frac{\pi+\theta}{\pi-\theta}\right)^{-\theta/\pi}$$
$$\times\left[\left(1+3\cos^2\theta\right)K_I^2-4\sin(2\theta)K_I K_{II}+\left(9-5\cos^2\theta\right)K_{II}^2\right], \tag{7}$$

where E is Young's modulus. According to the assumptions made by Palaniswamy and Knauss [14], the direction of crack propagation is the orientation in which the maximum rate of elastic energy release rate once exceeds a critical value; that is,

$$\frac{\partial G}{\partial \theta} = 0, \qquad \frac{\partial^2 G}{\partial \theta^2} < 0. \tag{8}$$

NT -Criterion

In NT-criterion [19] the total strain energy density N (equivalent to S in (4)) is split into the distortional stain energy density N_d and dilatational stain energy density N_v; that is,

$$N = N_d + N_v, \tag{9}$$

Where

$$N_d = \frac{1}{6\mu} \times \left(\sigma_{xx}^2 + \sigma_{yy}^2 + \sigma_{zz}^2 - \sigma_{xx}\sigma_{yy} - \sigma_{yy}\sigma_{zz} - \sigma_{zz}\sigma_{xx} + 3\sigma_{xy}^2\right),$$

$$N_v = \frac{1}{18K}\left(\sigma_{xx} + \sigma_{yy} + \sigma_{zz}\right)^2, \qquad K = \frac{E}{3(1-2\nu)}. \tag{10}$$

The elastic-plastic boundary then can be obtained by employing a suitable yield condition, and in NT-criterion, the Von Mises yield function is adopted. Then, one can have

$$N_{d,0} = \frac{\sigma_y^2}{6\mu}, \tag{11}$$

where σ_y is the critical material yield strength. NT-criterion postulates that a crack propagates in the direction of maximum dilatational strain energy $N_{V,max}$ along the contour of constant distortional strain energy $N_{d,0}$ when the distance from crack tip to the elastic-plastic boundary r_p in the growth direction reached a critical value r_{pcr}; that is,

$$\frac{\partial N_{v|r_p}}{\partial \theta} = 0, \qquad \frac{\partial^2 N_{v|r_p}}{\partial \theta^2} < 0, \tag{12}$$

Where $r_{pcr} = (1-2v)^2 K_{Ic}^2/(2\pi\sigma_y^2)$ for plane strain and $r_{pcr} = K_{Ic}^2/(2\pi\sigma_y^2)$ for plane stress case.

Y -Criterion

Y-criterion [20] also divides the total energy density into two components, the dilatational stain energy density Y_v and the distortional strain energy density Y_d, as

$$Y_d = N_d, \qquad Y_v = N_v. \tag{13}$$

Y-criterion presumes that a crack growth occurs in the direction of minimum distortional strain energy density along the constant core region boundary r_0 when the minimum distortional strain energy density Y_{d,θ_0}, reached a critical value $Y_{d,0}$, and it can expressed as

$$\frac{\partial Y_{d|r_0}}{\partial \theta} = 0, \qquad \frac{\partial^2 Y_{d|r_0}}{\partial \theta^2} > 0. \tag{14}$$

MIXED MODE FRACTURE TESTS ON DOLOMITE ROCK

Rock Properties

The experimental material, dolomite rock, was taken from a quarry in Fangshan County, Beijing. Visual inspections indicated that the rock was isotropic, homogenous, and white in color. Its structure was continuous and very tight. The X-ray diffraction (XRD) analysis (Figure 1) showed that the rock was composed of dolomite ($CaMg(CO_3)_2$) and quartz (SiO_2). No pores or microcracks were visible using SEM with ×50 magnification (Figure 2), therefore, the porosity of this rock can be neglected. SEM images also identified the material as a microcrystalline-textured, fine grained rock. Its dry density was about 2.59 g/mm^3. In order to get the mechanical properties of dolomite rock, 4 groups of uniaxial compression tests and 4 groups of Brazilian tests were conducted. The mechanical properties of the rock included a uniaxial unconfined compressive strength of 135 MPa, a tensile strength of 3.0 MPa, Young's modulus of elasticity of 50 GPa, and Poisson's ratio of 0.25.

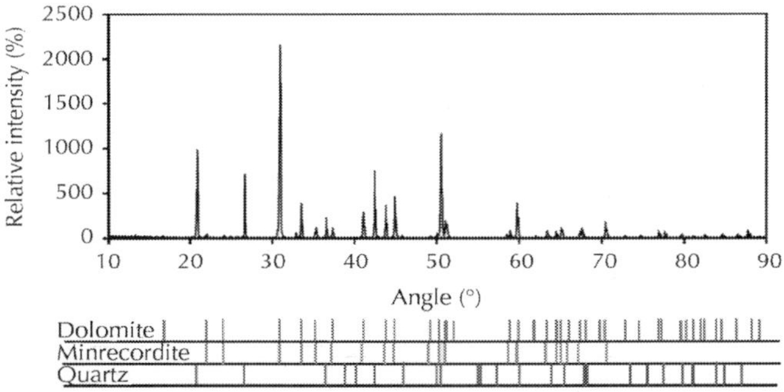

Figure 1: XRD results for dolomite rock.

Figure 2: SEM image for dolomite rock with ×50 magnifications.

Sample Preparation

The SCB specimens [21] were selected for mixed mode fracture tests on dolomite rock. Figure 3 shows the geometry of the SCB specimen and the loading setup of the experiments. SCB specimen was a half disc of radius R and thickness t .The edge crack was of length a and angle α relative to the loading direction. The specimen was loaded by a vertical load P under three-point bending and the distance between the bottom supports was 2s. When α was zero, the specimen was subjected to pure mode I (opening mode) independent of a/R and s/R [22]. By increasing the angle α, the contribution of mode II (or sliding mode) to the crack deformation increased. Pure mode II took place at a specific angle α depending on a/R and s/R [22].

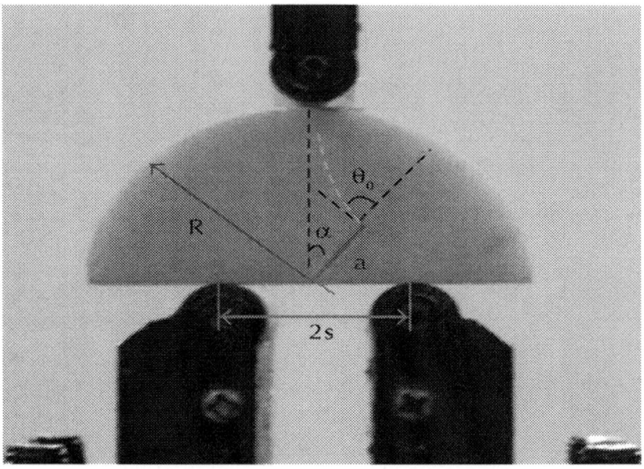

Figure 3: Geometrical characteristics of specimens and loading setup.

To avoid any possible boundary effects, the specimens were taken from the center of the rocks. The SCB specimens had a radius of 25 mm and were 15 mm in thickness. Values of a/R=0.4 and s/R=0.4 were selected for all the specimens. The cracks were created using a 0.2 mm thick diamond impregnated wire saw and the crack length a was 10 mm. According to [22] and numerical analysis, for the selected specimen configuration and loading condition, at an inclination angle of 34.8° the crack was pure mode II crack. Therefore, the crack inclination angles of α = 0° (pure mode I), 10°, 20°, 30°, 34.8° (pure mode II), and 40° (compression-shear mode) were selected. For each crack angle, four samples were prepared.

Fracture Tests

Each SCB specimen was located inside a three-point bending fixture (Figure 3) and then loaded by a test machine with a capacity of 10 kN. The loading rate was set to 0.1 mm/min for all the experiments. Figure

4represents a typical plot of the applied load against load-point displacement. It can be seen that after a nonlinear stage the load-displacement curve gives a linear relationship up to the maximum load. The dolomite specimens, including the crack, remain stable without any visible changes during the loading process until the load reaches a critical value. At that point the load-displacement curve drops suddenly, which is the characteristic of brittle fracture; the original crack propagates rapidly toward to the point where the load is applied. When the crack reaches the edge of the specimen, the specimen breaks into two pieces (Figure 5). The fracture surfaces can reveal some useful information about the failure mechanism. Therefore, SEM was used to conduct a microscale investigation on the typical fracture surfaces.

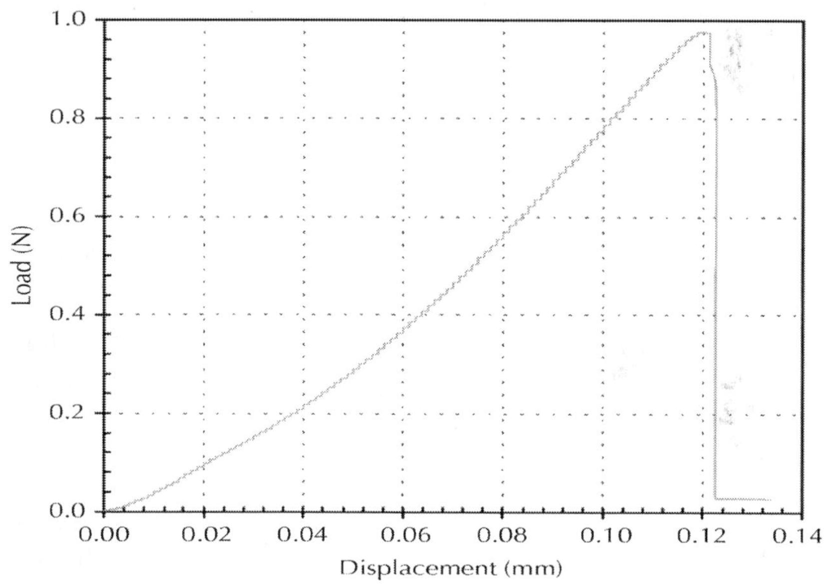

Figure 4: Typical plot of the applied load against the load-point displacement.

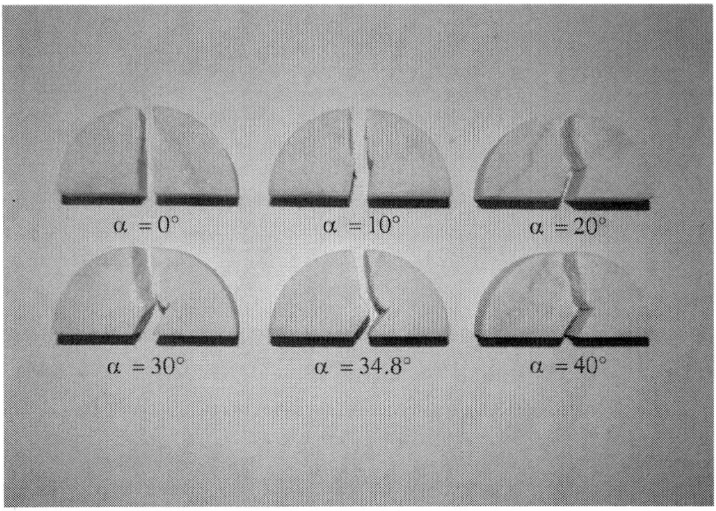

Figure 5: SCB specimens after failure.

Test Results and Analysis

The failure loads p_{cr} obtained from the mixed mode fracture tests are listed in Table 1. The average values of p_{cr} for α= 0°, 10°, 20°, 30°, 34.8°, and 40° are 840 N, 918 N, 930 N, 982 N, 1034 N, and 1127 N, respectively. The failure load increases gradually as the proportion of mode II deformation increases.

Table 1: Failure loads p_{cr} obtained from mixed mode fracture tests

Specimen code	**Crack inclined angle (°)**	**Failure load (N)**
D-1 (pure mode I)	0	832
D-2 (pure mode I)	0	857
D-3 (pure mode I)	0	846
D-4 (pure mode I)	0	836
D-5	10	925

D-6	10	910
D-7	10	892
D-8	10	945
D-9	20	960
D-10	20	810
D-11	20	929
D-12	20	901
D-13	30	1002
D-14	30	936
D-15	30	997
D-16	30	993
D-17 (pure mode II)	34.8	1001
D-18 (pure mode II)	34.8	1035
D-19 (pure mode II)	34.8	1210
D-20 (pure mode II)	34.8	890
D-21	40	956
D-22	40	1256
D-23	40	1254
D-24	40	1042

Crack initiation includes two components: the crack initiation angle and the crack initiation point. According to linear elastic fracture mechanics (LEFM), the crack initiation point is usually expected to be the crack tip of the original crack. However, for specimens with α=30° and 34.8° (pure mode II), the crack initiation point is not from the crack tip, as shown in Figure 5. Generally, only when the crack inclination angle α is very large, usually bigger than the values of pure mode II or when the crack tip is obtuse seriously, the crack initiation point may not originate from the crack tip. However, the two cases in the tests do not fall into either of these two categories. In order to better explain such phenomenon, the fracture surface near the crack tip has been observed with SEM. It is found that the mineral grain near the crack tip is much larger than that in the surrounding area (Figure 6), which may lead to this unexpected failure behavior. The other reason may be that the specimen contains quartz, which possesses a much higher strength than dolomite. It is difficult to evaluate the influence of the size of the particles and the mineral composition on the crack initiation, either qualitatively or quantitatively.

Figure 6: Size of particles around the crack tip (α=30°).

Dolomite is a brittle rock with a relatively low hardness. After the failure of specimens, an obvious damaged zone can be found for each specimen, as illustrated in Figure 7. The damage zone is narrow relative to the specimen. The color of the damaged zone is different from the surrounding area; its color is much whiter obviously. Regardless of the position of crack tip, a high stress zone must exist around the crack tip and lead to damage. Fonseka et al. [23] has confirmed this view using the technology of SEM and acoustic emission (AE). But this is a macroscopic experiment that may demonstrate directly that the emergence of a macroscopic crack is the result of damage that develops from microcracks and microvoids.

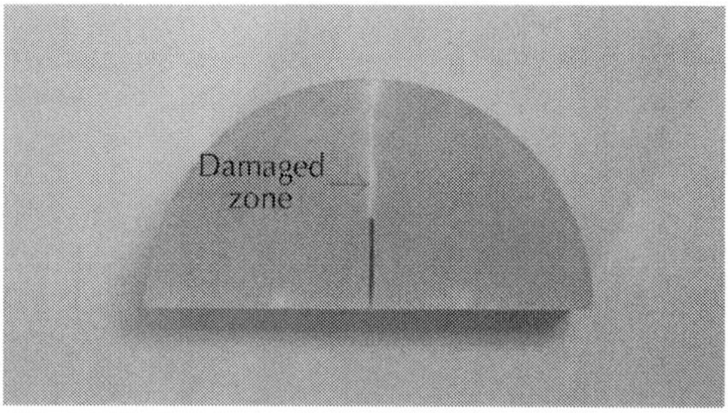

Figure 7: Visible damaged zone of a specimen.

SEM image was used to examine the failure manner to better understand the fracture mechanism and the evolution of the damage in dolomite rock. The fracture faces near the original crack tip for the cases of α= 0°, 10°, 20°, 30°, 34.8°, and 40° were observed by SEM; the fracture faces produced by the Brazilian test were also studied for comparison. Figure 8 shows SEM images of crack surface of the dolomite SCB specimens and Brazilian test with ×100 magnification.

(a)

(b)°

(c)°

(d)°

(e)°

(f)

(g)

Figure 8: SEM images of fractured surfaces of SCB and Brazilian test specimens with ×100 magnification.

In the case of $\alpha=0°$ (Figure 8(a)), the cracked surface is coarse which was mainly caused by intergranular failure. Compared with the fracture surface of the broken sample from a Brazilian test (Figure 8(g)), the two faces are very similar, indicating that the specimen is destroyed by splitting and stretching. The SEM image for $\alpha=10°$ (Figure 8(b)) shows that the cracked surface is also produced primarily by splitting, tensile stress is the main factor, and only few of the crystalline particles are sheared off. There is no evidence of crushing or friction. The micrograph for the case of $\alpha=20°$ illustrates that the fractured surface is similar to that of image. The notable difference is the increment of transgranular damage, revealing that the impact of shear stress is increasing, although it is still small relative to the effect of splitting. There is still no evidence of crushing in the specimen of $\alpha=20°$. In the SEM image for $\alpha=30°$ the transgranular damage becomes more apparent; the fractured surface is produced by a combination of splitting and sliding shear. When the angles are $\alpha=34.8°$, and 40°, the SEM images show obvious crushing between the boundary of the crystals. For =34.8°, the failure is due to shear stress that is consistent with LEFM. For the specimens of $\alpha=40°$, crushing is more serious. The analysis indicates that when the angle increases from pure mode I to pure mode II, the effect of the shear stress grows, but there is no crushing or abrasion on the cracked surface, which is different from the results proposed in previous studies [9]. Moreover, when the angle reaches pure mode II or continues to increase (compression-shear mode), the signs of crushing become apparent, with many rock fragments observed on the fractured surface.

NUMERICAL ANALYSIS AND COMPARISON

Based on (1)–(14), the crack initiation direction and growth path can be derived. Equations (1)–(14) are functions of the stress intensity factors and various material constants. By using the conventional remesh method based on FEM, (1)–(14) were employed to theoretically predict the fracture paths. No dynamic effects were considered in the crack propagation process. The finite element code ABAQUS and the mathematical software MATLAB were combined to evaluate the direction of crack growth. Stress intensity factors calculated by ABAQUS would output to a specified file, and fracture criterion was written as a program by MATLAB. The program would read stress intensity factors from the specified file and calculate the angle in which the crack would propagate. In each step, the simulation model was remodeled with a small crack extension of 0.5–1 mm. In the initial steps, the extension was relatively small. As the crack tip gradually approached the loading point, the crack extension of each step became larger. Generally, simulating each model required at least 21 steps. The following procedure gave more details for predicting the crack evolution path using the five fracture criteria:

- calculating K_I and K_{II} for the existing crack,
- using each criterion to evaluate the corresponding crack growth direction,
- remodeling the crack with a small extension along the direction which was calculated in step 2,
- repeating steps 1–3 for the new crack growth.

The loading conditions were considered similar to those of the fracture tests; for α=0°, 10°, 20°, 30°, 34.8°, and 40°, the loads were 840 N, 918 N, 930 N, 982 N, 1034 N, and 1127 N, respectively. ABAQUS provides a domain integral method to compute the stress intensity factors directly for a crack under mixed-mode loading. To produce the square root singular of the stress/strain field, the singular elements were considered in the first ring of elements surrounding the crack tip. In order to ensure a smooth curve which was the boundary of domain integral used to calculate stress intensity factors, 8 rings of quadratic quadrilateral elements surrounding the singular elements were meshed by the meshing technique of sweep. Other regions of the numerical model were meshed by the meshing methods of sweep or free, and the element type was also quadratic quadrilateral elements. A total of about 5000 eight-node plane strain elements (CPE8) were used for each model. Figure 9 represents a typical finite element mesh used for numerical analysis. The alterations of K_I and K_{II} with the mesh density are shown in Figure 10. It is observed that the values K_I and K_{II} are not sensitive to mesh density. The mesh density adopted in this research is reasonable. Using the method described above, the crack initiation angle and crack propagation path for different crack inclined angles were analyzed. The experimental and numerical results were compared to evaluate the applicability of the fracture criteria to predict the crack evolution path of dolomite rock.

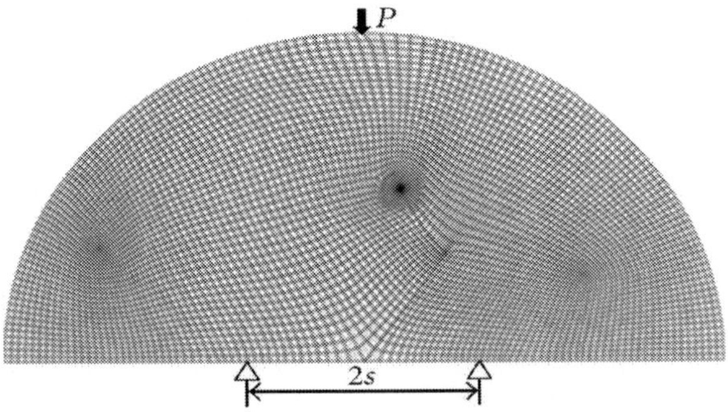

Figure 9: Typical finite element mesh used for numerical analysis.

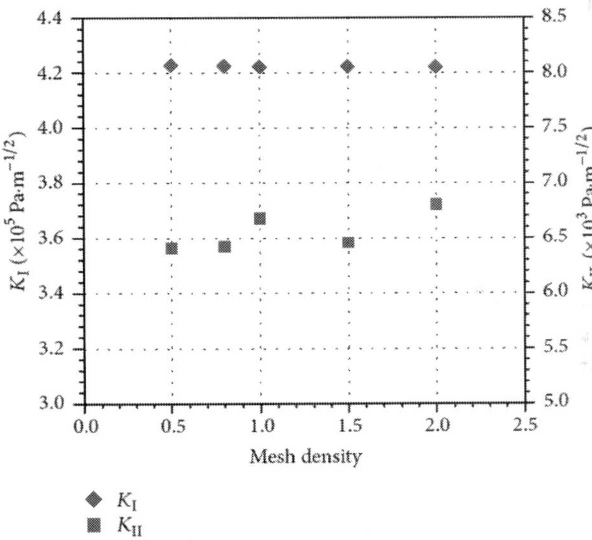

Figure 10: Alterations of K_I and K_{II} with mesh density (the mesh density adopted (Figure9 in the paper) for numerical simulation in this research is assumed to be 1; the mesh density of other numerical models is 0.5, 0.75, 1.5, and 2 times as dense as Figure 9).

The impact of Poisson's ratio on the crack initial direction and crack propagation path was also considered. Poisson's ratio of the experimental dolomite SCB specimens was 0.25. Thus numerical models with Poisson's ratio of 0.1, 0.25, and 0.4 were simulated.

Crack Initiation Angle

The crack initiation angle θ_0 is the direction in which the crack propagates from the original crack (Figure 3). For different α, the value of θ_0 is calculated by each fracture criteria introduced in Section 2. Table 2summarizes the crack initiation angle from simulated results.

Table 2: Summary of the crack initiation angle θ_0 of simulation for different crack inclination angle α (°)

α	σ_{θ}-criterion	S_{min}-criterion	G_{max}-criterion	NT-criterion	Y-criterion
0	0	0	0	0	0
10	29.1	26.0	29.8	29.3	29.0
20	48.8	44.7	51.6	52.8	51.1
30	63.7	66.5	68.5	81.3	74.8
34.8	70.5	80.4	75.2	98.9	88.4
40	70.5	80.4	75.2	98.9	88.4

Generally, the point possesses maximum stress, and maximum energy is at different locations and in a different direction, so the crack initiation angle predicted by different fracture criteria should be different. When mode I deformation is dominant (α<10°), the difference in crack initiation angle is very small. The maximum difference between the predicted angles is 3.8° (between the S_{min} -criterion and G_{max} -criterion), and there is little difference between the σ_θ -criterion and Y-criterion. Hence, when the angle α is small, that is, the proportion of the mode II deformation is relatively small, the difference of the simulated results by the five fracture criteria is negligible. As α increases, the difference in initiation angle between the various fracture criteria becomes apparent. When α=20°, the difference between different criterions becomes larger. The predicted angle for the G_{max} -criterion is 51.6° and for the S_{max} -criterion is 44.7°. The difference is about 7°. For =30°, the biggest difference between different criterions is nearly 18°. When α=34.8°, the predicted value changes from 70.5 to 98.9°, a difference of almost 30°. The proportion of the mode II deformation plays an important role in the predicted results of different fracture criteria. When α=40°, the crack belongs to compression crack. From the perspective of physics, the materials on both sides of the closed crack surface cannot penetrate into each other, and therefore, such cracks are pure model II cracks according to the LEFM. Thus, the initiation angle for the specimen of α=40° would be the same as the case of α=34.8°, as shown in the last two rows of Table 2. However, this is inconsistent with the experimental results. For α being larger than the angle of pure mode II, the fracture criteria are not applicable.

Crack Propagation Path

The dolomite specimens' failure paths under various combination loadings of mode I and mode II are plotted in Figure 11. When the sample is subjected to pure mode I, the crack propagates straight along the line of the original crack. As the effect of mode II deformation increases, the crack evolution path becomes curved. When α=40°, the crack path is the most curving. Figure 12 shows the simulation results of crack trajectories for each inclined angle specimens using the five fracture criteria.

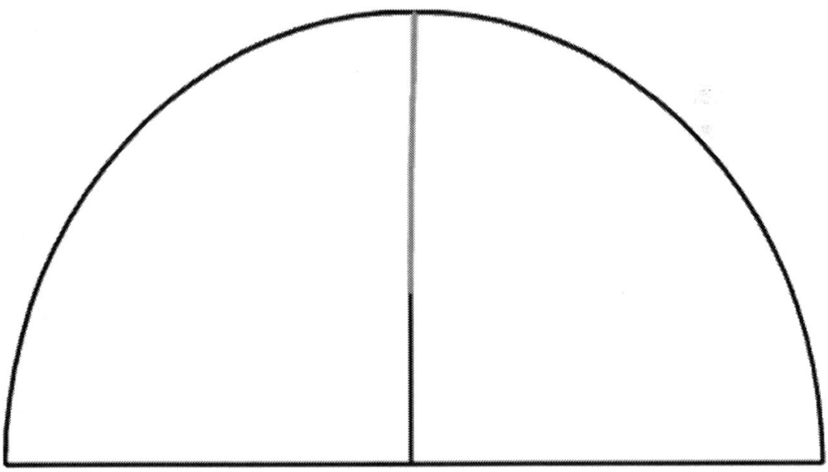

(a)

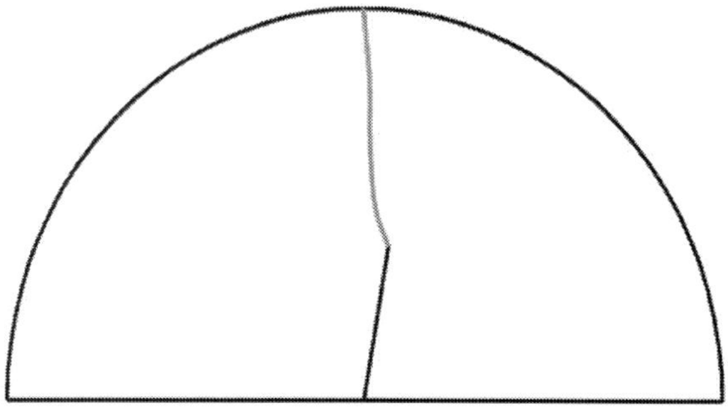

(b)

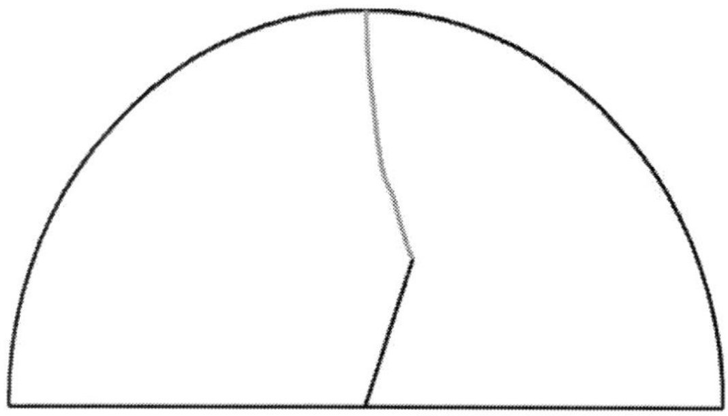

(c)

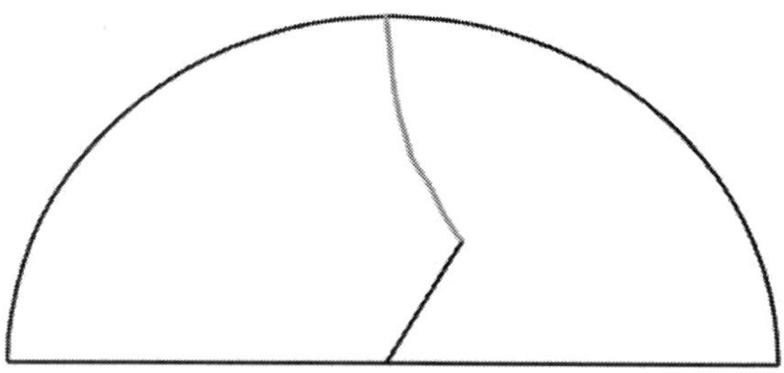

(d)

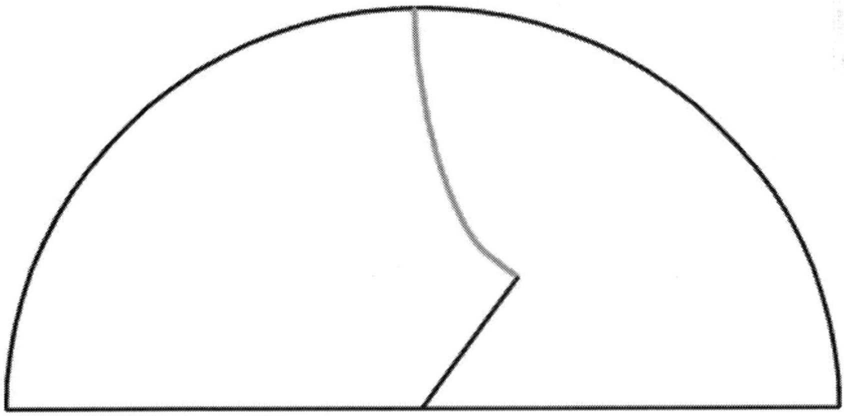

(e)

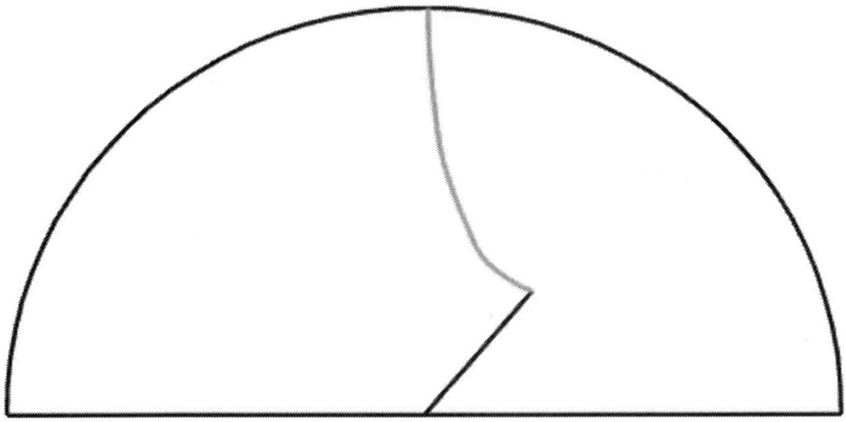

(f)

Figure 11: Crack path of fractured SCB specimens with different crack inclination angle α.

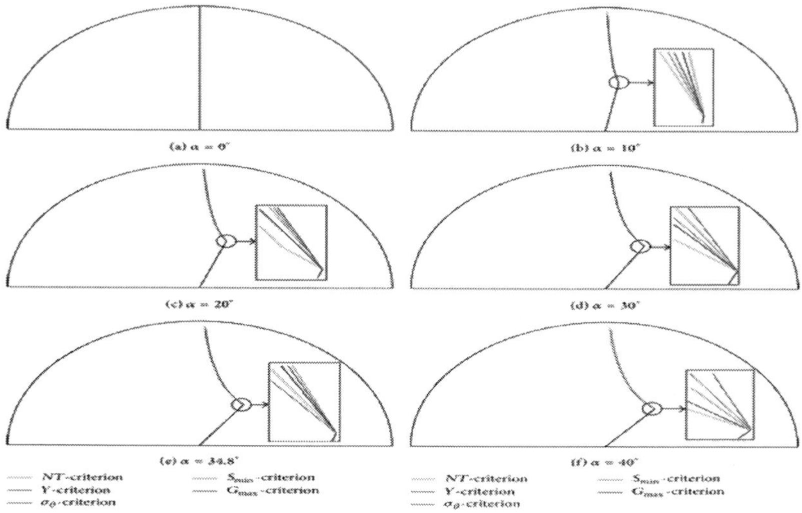

Figure 12: Simulated results of crack trajectories. The red, green, blue, magenta, and cyan lines represent the paths predicted by the σ_θ-criterion, S_{min}-criterion, Y-criterion, G_{max}-criterion, and NT-criterion, respectively.

When =0°, the crack extension angle is 0° for each fracture criterion (Figure 12(a)). As increases from 10° to 40°, the simulated crack trajectories are very similar for all the fracture criteria. To further examine the crack growth paths calculated by different criteria, the plots of the crack trajectories near the original crack tip are enlarged. At the initial part of crack growth path, the difference between the various fracture criteria is relatively large. This is due to the initial angle of the crack varies between different fracture criteria, but after a few steps of crack growth, the difference between the predicted crack trajectories of the various fracture criteria become smaller and can be neglected. Therefore, any one of the five simulated results can be used for comparison with the experimental results.

After the initial stage of crack propagating, mode I deformation becomes dominant. Taking =34.8° (pure mode II) as an example, after four steps of simulation, the stress intensity factors K_I and K_{II} calculated by the S_{min}-criterion are 422200 and 6676 $Pa \cdot m^{-1/2}$, respectively; for the NT-criterion, Y-criterion, G_{max}-criterion, and σ_θ-criterion, the values of K_I and K_{II} are 437400 and 10280 Pa·m$^{-1/2}$, 430400 Pa·m$^{-1/2}$ and 8607 Pa·m$^{-1/2}$, 427500 and 3186 Pa·m$^{-1/2}$, and 429500 and 5892 Pa·m$^{-1/2}$, respectively. It is obvious that the proportion of mode II deformation is relatively small compared with that of mode I. Based on the discussion in Section 4.1, when the proportion of mode I deformation is dominant, the estimated crack initiation angles of all five criteria are similar, which leads to the simulated crack trajectories of the various fracture criteria are similar when the new crack tip moves slightly away from the original crack tip. Examining the predicted crack trajectories for the different angles of inclination using one fracture criterion, and NT-criterion is taken as an example (Figure 13); it is found that when the crack extends to the loading point, the crack paths gradually move close to each other; the same reasons mentioned above can explain this phenomenon.

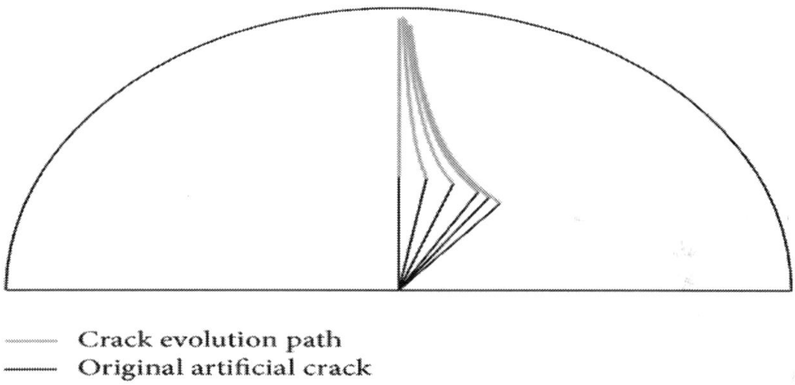

Figure 13: Crack trajectories for the different angles of inclination predicted by one fracture criterion (take NT-criterion as an example).

Comparison between the Experimental and Simulated Results

The crack trajectories predicted based on the five fracture criteria are very similar. Therefore, for simplification, only the simulated crack evolution paths of the S_{min}-criterion were selected to compare with the experimental results. Figure 14 describes the experimental results (crack evolution paths and damaged zones) and the calculated trajectories based on the S_{min}-criterion.

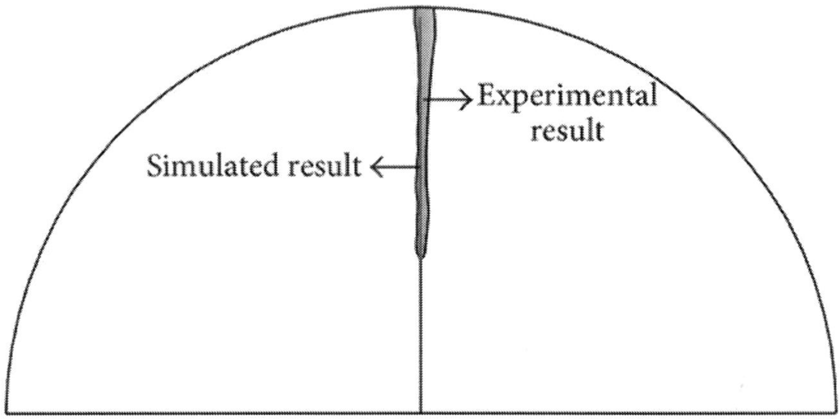

Figure 14: Experimental crack propagated paths, damaged zones, and crack trajectories calculated by S_{min}-criterion (the gray zone, red line, and green line represent the damaged zone, numerical result, and experimental result, resp.).

For cracks of pure mode I, the test result coincides with the simulation result of the fracture criteria. When α=10°, the simulation paths are also in good agreement with the experimental results. For =20°, the difference between the experimental result and the simulation result becomes larger but still acceptable. It is interesting that all the predicted crack trajectories of α=0°, 10°, and 20° are in the damaged zone. That is, the appearance of crack is the consequence of damage evolution and the crack propagation path should locate in the damaged zone. Therefore, when mode I deformation is predominant, the results predicted by the five fracture criteria are reliable. In current work, for predominately mode II cracks (α=30°, 34.8°, and 40° (Figures 14(d)–14(f))), the predicted crack paths locate outside the damaged zones and are not consistent with the experimental results. For a predominately mode II deformation, all the fracture criteria fail to predict the crack path.

Influence of Poisson's Ratio

Equations (6), (12), and (14) indicate that Poisson's ratio v has an influence on the numerical analysis results. To assess the effect of v quantitatively, three different values (v=0.1, 0.25, and 0.4) were selected for the numerical analysis. Table 3 summarizes the crack initiation angles θ_0 for different Poisson's ratio calculated by the S_{min}-criterion, NT-criterion, and Y-criterion. The last two rows of Table 3 are the same and the reasons have been explained in Section 4.1. For the NT-criterion, the crack initiation angles do not change with Poisson's ratio, indicating that Poisson's ratio has no effect on the NT-criterion prediction of the crack initiation angle. For the S_{min}-criterion, as Poisson's ratio increases, the predicted crack initiation angle becomes larger. When $\alpha<10°$, the difference is relatively small; the maximum difference is less than 7°. For $\alpha<10°$, the difference between the results of v=0.1 and v=0.4 is more than 11°. Poisson's ratio plays an important role in the predicted angle of the S_{min}-criterion. The effect of Poisson's ratio on the Y-criterion is similar to that of the S_{min}-criterion, but the effect is much smaller. For v=0.1 and v=0.4, the biggest difference for the Y-criterion is less than 5°. Poisson's ratio has the greatest influence on the S_{min}-criterion, followed by the Y-criterion, and no effect on the NT-criterion. Furthermore, the crack trajectories are not sensitive to Poisson's ratio for the three fracture criteria. The simulated results of the crack evolution path are similar to those of Figures 12(a)–12(e).

Table 3: Summary of the crack initiation angles θ_0 for different Poisson's ratio (°)

S_{min}-criterion				NT-criterion			Y-criterion		
	v=0.1	v=0.25	V=0.4	v=0.1	v=0.25	v=0.4	v=0.1	v=0.25	v=0.4
0	0	0	0	0	0	0	0	0	0
10	22.6	26.0	28.6	29.3	29.3	29.3	28.6	29.0	29.3
20	37.4	44.7	49.6	52.8	52.8	52.8	49.4	51.1	51.9
30	59.5	66.5	72.6	81.3	81.3	81.3	72.4	74.8	76.1
34.8	74.5	80.4	86.2	98.9	98.9	98.9	85.9	88.4	89.8
40	74.5	80.4	86.2	98.9	98.9	98.9	85.9	88.4	89.8

CONCLUSIONS

Using SCB specimens of dolomite rock, mixed mode brittle fracture was studied by experiments. In order to explain the observed failure behavior, the SEM technique was used to study the microfailure mechanism. Furthermore, by using the conventional remesh method based on FEM, the crack growth trajectories for various mixed mode loading conditions, from pure mode I to mode II, were theoretically (by the σ_θ-criterion, S_{min}-criterion, G_{max}-criterion, NT-criterion, and Y-criterion) and numerically predicted. The applicability of the five criteria was examined. Finally, an investigation on the effect of Poisson's ratio on fracture criteria was also presented. Based on the current work, the following conclusions can be obtained.

- The dolomite specimens fracture in a brittle manner: as the proportion of mode II deformation increased, the average failure load increased.
- Damaged zones were observed on both sides of the crack. It was a macroscopic experiment that could demonstrate that the emergence of macroscopic cracks was the result of the evolution of damage.
- When the crack inclination angle increased, the effect of the shear stress increased but no crushing or abrasion was observed on the cracked surface. When the crack inclined angle was 34.8° or 40°, evidences of crushing between the cracked faces began to appear.
- When mode I deformation was dominant, the difference between the crack initiation angles predicted by the five fracture criteria was small. As the proportion of mode II deformation increased, the difference became larger. Mode II deformation had a considerable impact on the crack initiation angle.
- The crack trajectories predicted by the five fracture criteria were very similar and the differences among the prediction paths were negligible.
- The predicted crack trajectories were located in the damaged zones for predominately mode I cracks and in such cases, the fracture criteria were suitable; when mode II deformation was

predominant, all five fracture criteria failed to predict the crack trajectories.

- The crack initiation angle predicted by the Y-criterion and S_{min}-criterion was sensitive to Poisson' ratio, particularly the S_{min}-criterion; however, the crack trajectories were not sensitive to Poisson's ratio for these two fracture criteria. For the NT-criterion, the crack initiation angle and the crack trajectories were independent of Poisson's ratio.

ACKNOWLEDGMENTS

This work was financially supported by the Provincial Science and Technology Support Project (2012FZ0124), Major State Basic Research Project (2011CB201201), and the International Science & Technology Cooperation Program of China (2012DFA60760).

REFERENCES

1. S.-H. Chang, C.-I. Lee, and S. Jeon, "Measurement of rock fracture toughness under modes I and II and mixed-mode conditions by using disc-type specimens," Engineering Geology, vol. 66, no. 1-2, pp. 79–97, 2002.
2. I. L. Lim, I. W. Johnston, S. K. Choi, and J. N. Boland, "Fracture testing of a soft rock with semi-circular specimens under three-point bending. Part 2-mixed-mode," International Journal of Rock Mechanics and Mining Sciences and, vol. 31, no. 3, pp. 199–212, 1994.
3. Z. Jia, A. Castro-Montero, and S. P. Shah, "Observation of mixed mode fracture with center notched disk specimens," Cement and Concrete Research, vol. 26, no. 1, pp. 125–137, 1996.
4. G. S. Xeidakis, I. S. Samaras, D. A. Zacharopoulos, and G. E. Papakaliatakis, "Trajectories of unstably growing cracks in mixed mode I-II loading of marble beams," Rock Mechanics and Rock Engineering, vol. 30, no. 1, pp. 19–33, 1997.
5. A. Bobet and H. H. Einstein, "Fracture coalescence in rock-type materials under uniaxial and biaxial compression," International

Journal of Rock Mechanics and Mining Sciences, vol. 35, no. 7, pp. 863–888, 1998.

6. C.-S. Chen, E. Pan, and B. Amadei, "Fracture mechanics analysis of cracked discs of anisotropic rock using the boundary element method," International Journal of Rock Mechanics and Mining Sciences, vol. 35, no. 2, pp. 195–218, 1998.
7. B. Vásárhelyi and A. Bobet, "Modeling of crack initiation, propagation and coalescence in uniaxial compression," Rock Mechanics and Rock Engineering, vol. 33, no. 2, pp. 119–139, 2000.
8. R. H. C. Wong, C. A. Tang, K. T. Chau, and P. Lin, "Splitting failure in brittle rocks containing pre-existing flaws under uniaxial compression," Engineering Fracture Mechanics, vol. 69, no. 17, pp. 1853–1871, 2002.
9. N. A. Al-Shayea, "Crack propagation trajectories for rocks under mixed mode I-II fracture," Engineering Geology, vol. 81, no. 1, pp. 84–97, 2005.
10. H. Y. Liu, S. Q. Kou, P.-A. Lindqvist, and C. A. Tang, "Numerical modelling of the heterogeneous rock fracture process using various test techniques," Rock Mechanics and Rock Engineering, vol. 40, no. 2, pp. 107–144, 2007.
11. M. R. M. Aliha, M. R. Ayatollah, D. J. Smith, and M. J. Pavier, "Geometry and size effects on fracture trajectory in a limestone rock under mixed mode loading," Engineering Fracture Mechanics, vol. 77, no. 11, pp. 2200–2212, 2010.
12. F. Erdogan and G. C. Sih, "On the crack extension in plates under plane loading and transverse shear,"Journal of Basic Engineering, vol. 85, no. 4, pp. 519–527, 1963.
13. G. C. Sih, "Strain-energy-density factor applied to mixed mode crack problems," International Journal of Fracture, vol. 10, no. 3, pp. 305–321, 1974.
14. K. Palaniswamy and W. G. Knauss, "Propagation of a crack under general, in-plane tension,"International Journal of Fracture Mechanics, vol. 8, no. 1, pp. 114–117, 1972.
15. M. A. Hussain, S. L. Pu, and J. Underwood, "Strain energy release rate for a crack under combined Mode I and Mode II," Fracture Analysis STP 560, ASTM, 1974.

16. R. J. Nuismer, "An energy release rate criterion for mixed mode fracture," International Journal of Fracture, vol. 11, no. 2, pp. 245–250, 1975.
17. C. H. Wu, "Fracture under combined loads by maximum-energy-release-rate criterion," Journal of Applied Mechanics—Transactions ASME, vol. 45, no. 3, pp. 553–558, 1978.
18. A. P. Kfouri and M. W. Brown, "Fracture criterion for cracks under mixed-mode loading," Fatigue & Fracture of Engineering Materials and Structures, vol. 18, no. 9, pp. 959–969, 1995.
19. N. A. B. Yehia and M. S. Shephard, "The nt-criterion for predicting crack growth increments,"Engineering Fracture Mechanics, vol. 26, no. 3, pp. 371–382, 1987.
20. N. A. B. Yehia, "Distortional strain energy density criterion: the Y-Criterion," Engineering Fracture Mechanics, vol. 39, no. 3, pp. 477–485, 1991.
21. K. P. Chong and M. D. Kuruppu, "New specimen for fracture toughness determination for rock and other materials," International Journal of Fracture, vol. 26, no. 2, pp. R59–R62, 1984.
22. M. R. Ayatollahi and M. R. M. Aliha, "Wide range data for crack tip parameters in two disc-type specimens under mixed mode loading," Computational Materials Science, vol. 38, no. 4, pp. 660–670, 2007.
23. G. M. Fonseka, S. A. F. Murrell, and P. Barnes, "Scanning electron microscope and acoustic emission studies of crack development in rocks," International Journal of Rock Mechanics and Mining Sciences and Geomechanics Abstracts, vol. 22, no. 5, pp. 273–289, 1985.

Chapter 2

Fractal Analysis of Permeability of Unsaturated Fractured Rocks

Guoping Jiang[1], Wei Shi[2], and Lili Huang[2]

[1]Earthquake Engineering Research Test Center, Guangzhou University, Guangzhou 510405, China

[2]Ningbo Polytechnic, Ningbo, Zhejiang 315800, China

ABSTRACT

A physical conceptual model for water retention in fractured rocks is derived while taking into account the effect of pore size distribution and tortuosity of capillaries. The formula of calculating relative hydraulic conductivity of fractured rock is given based on fractal theory. It is an issue to choose an appropriate capillary pressure-saturation curve in the research of unsaturated fractured mass. The geometric pattern of the fracture bulk is described based on the fractal distribution of tortuosity. The resulting water content expression is then used to estimate the unsaturated hydraulic conductivity of the fractured medium based on the well-known model of Burdine. It is found that for large enough ranges of fracture apertures the new constitutive model converges to the empirical Brooks-Corey model.

INTRODUCTION

Modeling water flow in unsaturated fractured rocks has received considerable attention in the last two decades. One of the main reasons for focussing on the study of liquid flow in this type of media is that spontaneous capillary imbibition is an important fundamental phenomenon existing extensively in a variety of processes such as oil recovery, polymer composite manufacturing, soil science, and hydrology. The other reason is that deep disposal in crystalline rocks is considered to be an effective mean of isolating radioactive wastes from the biosphere. The study of basic transport processes has long been recognized because of which one can heighten the comprehensive understanding of physical phenomena, such as permeability [1, 2], heat transfer [3–5], and sorption [6, 7]. Many parameters such as the porosity, size of pore, and tortuosity of capillaries are very important for fluid flow in hard rocks. These parameters, however, are closely related to the geometric architecture of hard rocks. Cai et al. have analyzed the natural fractured trace maps representing a wide variety of scales, geological settings, and lithologies [8]. Cai and Yu have reported the density of different sized fault segments within the San Andreas fault zone which is fractal [9]. The distributions of fracture aperture and fracture spacing are self-similar over a well-defined range of apertures in the Cajon Pass scientific drill hole which was found by Barton and Zoback [10]. Based on the assumption that the fracture pattern is self-similar, the Sierpinski carpet was often employed to simulate porous media [11]. The conceptual constitutive model proposed by Wu and Yu [12] had used its fractal dimension to the parameters of the Brooks-Corey constitutive model [13] through the Sierpinski carpet. A Sierpinski space was also adopted to characterize the spatial distribution of a drainage network in the Gardon basin, France [14].

From the above review it is shown that a mechanistic model has not yet been established. In this paper, we attempt to develop a mechanistic model for unsaturated flow in fractured hard rocks based on the method using the specific fractal to describe fractured rock. The expressions of the proposed constitutive model are closed form and easy to evaluate. Another important feature, the tortuosity fractal dimension D_T, which affected other model parameters and should not be neglected while it was often neglected in the past investigation, is considered.

Now fractal theory has a wide variety of applications in sciences and engineering fields such as thermal science [15–19], fluid science [1, 2, 6], and industrial construction engineering [20]. For example, Moussa [2] has systematically investigated the transport of porous media based on fractal theory. The work [2] is open. Xiao et al. [3–5, 15, 16, 18, 19] have done much outstanding work on heat transfer of fluids by using fractal technique. In our work, we derive the analytical expressions for the relative hydraulic conductivity of fractured rock while taking into account the effect of pore size distribution based on the fractal geometry theory.

CONSTRUCTION OF THE FRACTAL MODEL

The model is presented in Figure 1. It has been shown that the cumulative size distribution of contact spots on engineering surfaces is similar to islands on earth and pores in porous rock which follows the fractal scaling law [2].

$$N\left(L \geq D_n\right) = \left(\frac{D_{n,\max}}{D_n}\right)^{d_f}, \tag{1}$$

where d_f is the fractal dimension for pores, D is diameter, L is the length scale, and N is the total number of pores whose sizes equal to and greater than D_n. The number of pores whose sizes range from D_n to $D + dD_n$ is

$$-dN = d_f D_{n,\max}^{d_f} D_n^{-(d_f+1)} dD_n \tag{2}$$

When water flow through the pores of porous rock, the capillaries may be tortuous these tortuous capillaries could be expressed by fractal equation [21]. Consider

$$L_a(D_n) = L_0^{D_T} D_n^{1-D_T},$$

(3)

Where D_T is the tortuosity fractal dimension and lies in the range $1 < D_T < 2$, which represents the extent of convoluted ness of capillary pathways for fluid flow through a medium. Note that for a straight capillary path $D_T = 1$ and a higher value of D_T corresponds to a highly tortuous capillary. Let the diameter of a capillary in the medium be D_n and let its tortuous length along the flow direction be $L_a(D_n)$. L_0 is representative length of channels. With a straight capillary, $L_a(D_n) = L_0$.The total volume of pores from $D_{n,\min}$ to $D_{n',\max}$ can be obtained from (1) as

$$\begin{aligned} v_{tol} &= \int_{D_{n,\min}}^{D_{n,\max}} \frac{\pi}{4} D_n^2 L_a(-dN) \\ &= \int_{D_{n,\min}}^{D_{n,\max}} \frac{\pi}{4} D_n^2 L_a d_f D_{n\cdot\max}^{d_f} D_n^{-(d_f+1)} dD_n \\ &= \frac{\pi}{4(3-d_f-D_T)} L_0^{D_T} d_f D_{n\cdot\max}^{3-D_T} \\ &\times \left(1-\left(\frac{D_{n,\min}}{D_{n,\max}}\right)^{(3-d_f-D_T)}\right), \end{aligned}$$

(4)

Where v_{tol} is the whole volume of the pores, (-dN) is given by (2).

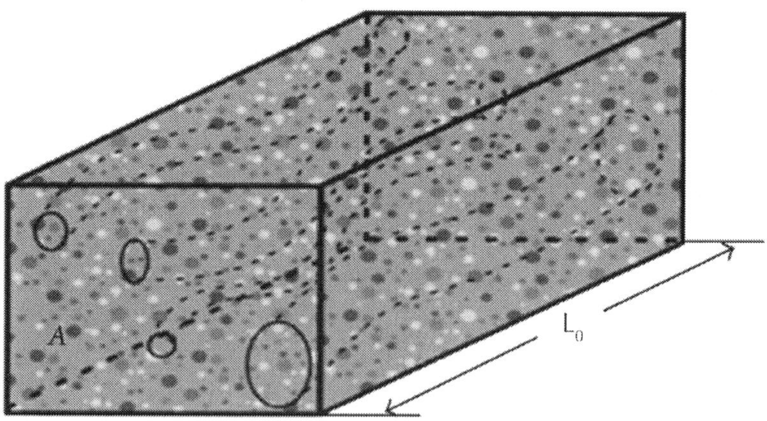

Figure 1: The model presented.

Similarity to the state above, the $v_{(D_{n,minx}<D_n)}$ volume of pores/particles from D_n to $D_{n,max}$ can be obtained from (1) as

$$v_{(D_{n,min}<D_n)}$$

$$= \frac{\pi}{4} L_0^{D_T} d_f D_{n\cdot max}^{d_f} \int_{D_{n,minx}}^{D_{n,}} D_n^{2-d_f-D_T} dD_n$$

$$= \frac{\pi}{4\left(3-d_f-D_T\right)} L_0^{D_T} d_f D_{n\cdot max}^{d_f} \left(D_n^{3-d_f-D_T}\right)\Big|_{D_{n,min}}^{D_{n,}} \tag{5}$$

The effective saturation S of volume $v_{(D_{n,minx}<D_n)}$ can be obtained. Consider

$$S = \frac{v_{(D_{n,min}<D_n)}}{v_{tol}}$$

$$= \frac{\Big|_{D_{n,min}}^{D_{n,max}} \left(D_n^{3-d_f-D_T}\right)}{\Big|_{D_{n,min}}^{D_{n,}} \left(D_n^{3-d_f-D_T}\right)}$$

$$= \frac{D_n^{3-d_f-D_T} - D_{n,min}^{3-d_f-D_T}}{D_{n,max}^{3-d_f-D_T} - D_{n,min}^{3-d_f-D_T}}. \tag{6}$$

Assuming the immiscible fluid flow in reservoir rocks, three relatively important forces are considered: the capillary pressure can be expressed. Consider

$$p = \frac{2\sigma\cos(\beta)}{\rho g D_n}, \tag{7}$$

where σ is surface tension between the wetting and non-wetting fluids, ρ is the water density, g is the gravity acceleration, D_n is diameter of a pore, and β is contact angle between the extraneous water and solid.

The relative permeability's are usually expressed in terms of water saturation S_w. The saturation curve for the proposed model of fractured rock was derived. Consider

$$S_w = \frac{p_c^{3-d_f-D_T} - p_{c\min}^{3-d_f-D_T}}{p_{c,\min}^{3-d_f-D_T} - p_{c,\max}^{3-d_f-D_T}}, \tag{8}$$

Where

$$p_{c\max} = \frac{2\sigma\cos(\beta)}{\rho g D_{n,\min}}, \qquad p_{c\min x} = \frac{2\sigma\cos(\beta)}{\rho g D_{n,\max}}. \tag{9}$$

THE PERMEABILITY OF UNSATURATED FRACTURED ROCKS

The Burdine and Mualem models are the two most widely used models predicting relative hydraulic conductivity. For the particular state of flow in fractured hard rock, the Burdine model seems to be more consistent and to be often adopted. The expression of the Burdine model is

$$K_D(s) = s^2 \frac{\int_0^s \left(ds/p_c^2\right)}{\int_0^1 \left(ds/p_c^2\right)}, \tag{10}$$

Where $K_D(s)$ is the relative hydraulic conductivity Inserting (8) to (10) we obtain the following form for $K_D(s)$

$$K_D(s) = \left(\frac{p_c^{d_f+D_T-3} - p_{c\max}^{d_f+D_T-3}}{p_{c,\min x}^{d_f+D_T-3} - p_{c,\max}^{d_f+D_T-3}} \right)^2 \times \frac{p_c^{d_f+D_T-5} - p_{c\max}^{d_f+D_T-5}}{p_{c\min x}^{d_f+D_T-5} - p_{c\max}^{d_f+D_T-5}}. \tag{11}$$

The novel constitutive model has some similarities with the well-known Brooks-Corey model, which is $S_0 = (p_c/p_d)^{-\lambda}$ and $K_D(s) = S^{3+\lambda/2}$, where p_d is the reciprocal of air entry pressure and λ is a model parameter related to pore size distribution. When $D_T = 1$, $\lambda = 2 - d_f$, and $p_{c,\max} \ll p_{c,\min x}$, $p_{c,\max}$ is neglected here, the model derived here is simplified to the Brooks-Corey model. Comparisons of the proposed and the Brooks-Corey models for three different ranges of fracture apertures are depicted in Figure 2. The assumed parameters are d_f = 1.8, $D_{n,\max} = 10^{-1}$ cm, and $D_{n,\min x} = 10^{-3}$ cm, 10^{-5} cm, and 10^{-7} cm. According to Figure 2, the Brooks- Corey model seems to be adequate to describe the hydraulic properties of fractured rocks for large ranges of fracture apertures and low values of pressure head.

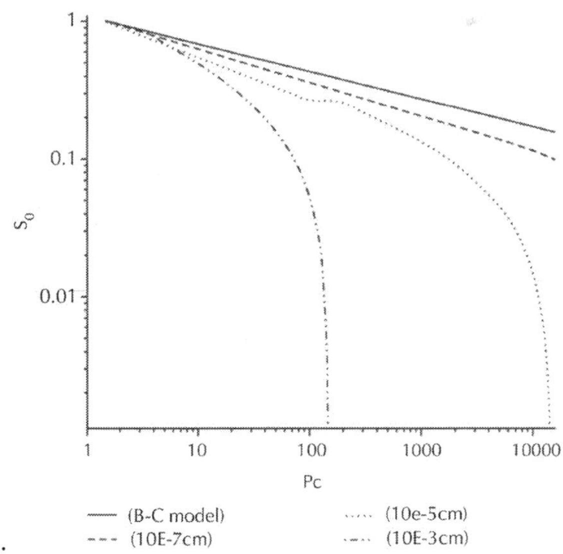

(a)

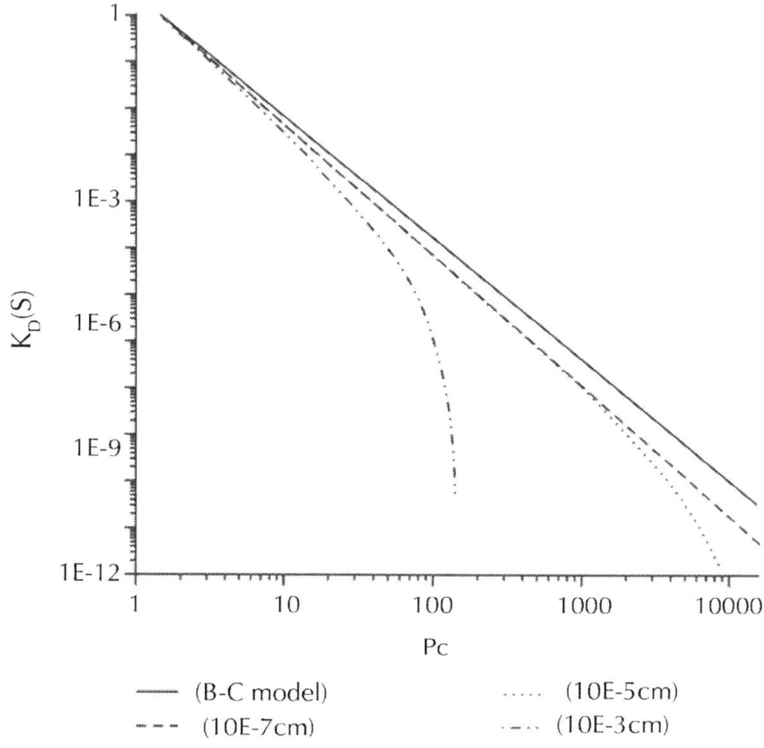

(b)

Figure 2: Comparison between the Brooks-Corey model and the new relations for different ranges of fracture apertures when D_T=1.

The relationship between the fractal dimension and effective saturation is very important in the study of unsaturated flow in fractured hard rocks. The geometric parameters and physical constants used for the analysis are $\beta = 0$, σ =72.25 dy/pa. In order to analyze the influence of the fracture density we consider different values of D_T. With (8) and (11), the relationship is obtained (Figure 3). With the physical constants D_n=0.1 cm, D_{nmax}=1 cm and $D_{nmin\ x}$=0.01cm, the relationship between d_f and s_0 can also be determined (Figure 4).

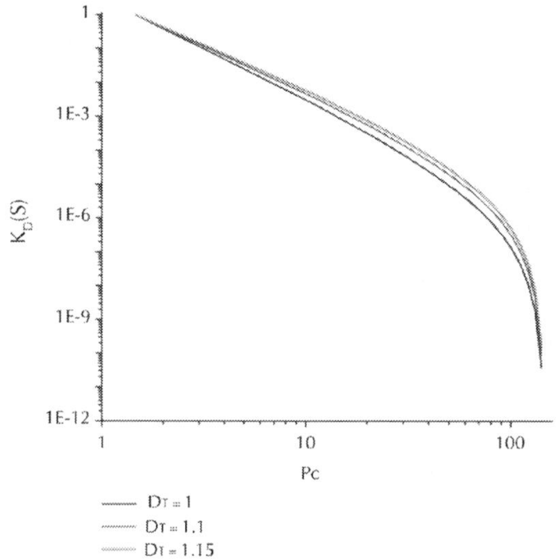

(a)

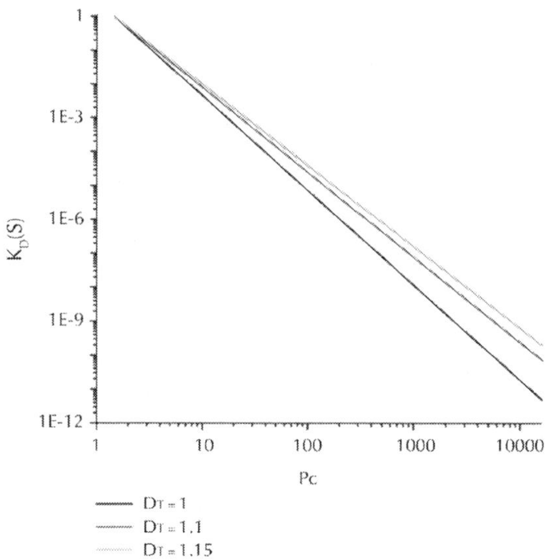

(b)

Figure 3: The new relations for K_D(s) and D_T.

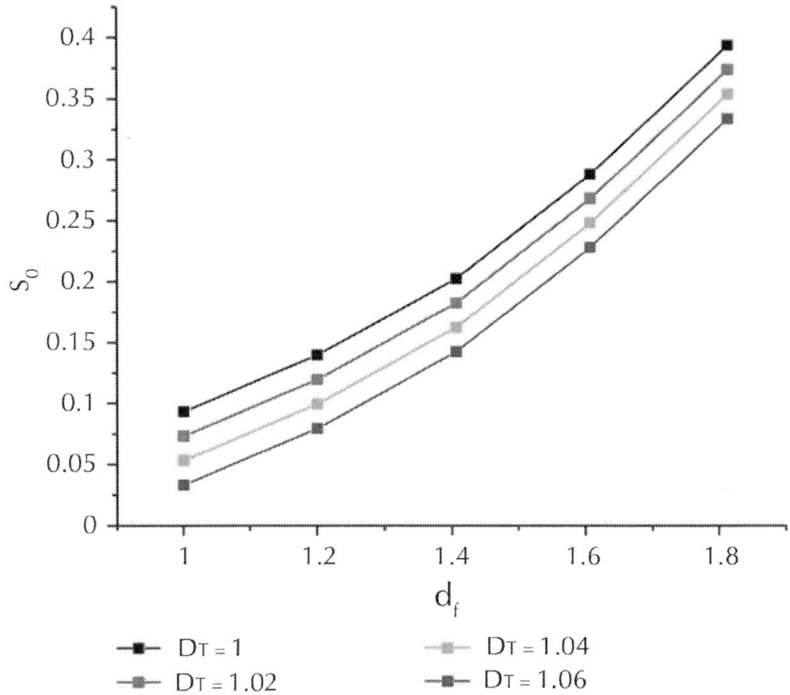

Figure 4: The new relations for S_0 and d_f.

In order to test the proposed invasion depth model, it is crucial to correctly determine the fractal dimensions D_T and g_f.According to the following formula proved recently by Yu and Li [22, 23],

$$\phi = \left(\frac{D_{n,\min}}{D_{n,\max}}\right)^{D_E - d_f}, \tag{12}$$

Where D_E is the Euclidean dimension and D_E=2 is used in this work. The maximum pore diameter can be calculated based on the model of square arrangement of particles.

According to the following formula proved recently by Jiangchao Cai, where K is permeability, K is Kozeny constant, which considers the tortuosity of capillaries and pore non-uniformity. And the parameters K and K are all proved [23]. Consider

$$D_T = 1 + \frac{\ln t_{av}}{\ln L_0/t_{av}},$$

$$D_{n,\max} = \frac{2(1-\phi)}{\phi}\sqrt{\frac{Kk}{\phi}}\left(\sqrt{\frac{\phi}{1-\phi}} + \sqrt{\frac{\pi}{4(1-\phi)}} - 1\right), \tag{13}$$

Where

$$t_{av} = \frac{1}{2}\left[1 + \frac{1}{2}\sqrt{1-\phi} + \frac{\sqrt{\left(\sqrt{1-\phi}-1\right)^2 + ((1-\phi)/4)}}{1-\sqrt{1-\phi}}\right],$$

$$\frac{L_0}{t_{av}} = \frac{d_f - 1}{2}\sqrt{\frac{1-\phi}{\phi}\frac{\pi}{d_f(2-d_f)}\frac{D_{n,\min}}{D_{n,\max}}}. \tag{14}$$

From (12) and (13), the relationship among ϕ, D_T, and d_f is determined (Figure 5).

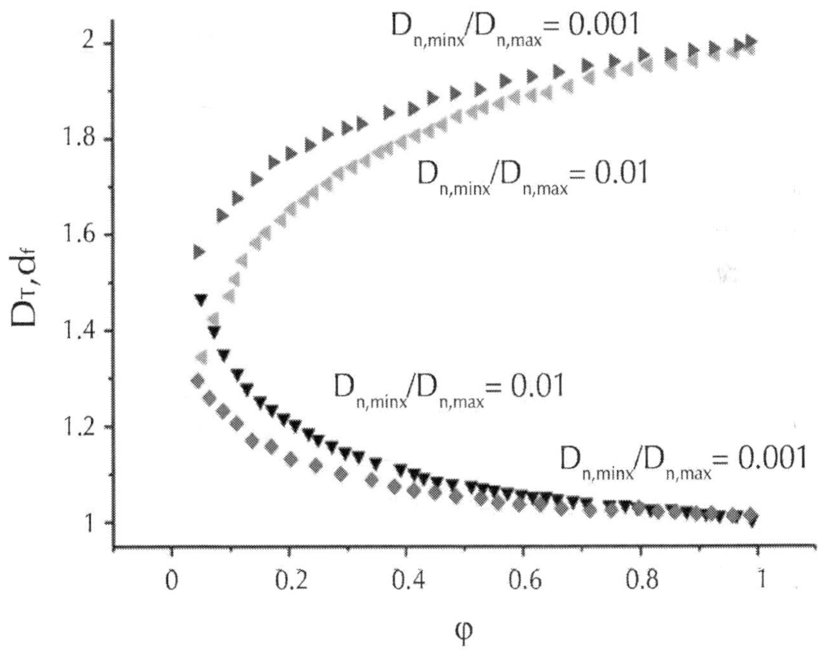

Figure 5: The relationship among ϕ, D_T, and d_f.

CONCLUSIONS

In this paper, with the consideration of pore size distribution and tortuosity of capillaries, a new fractal model for relative hydraulic conductivity of fractured rock is developed. The derived constitutive model is an effort to understand and characterize unsaturated flow in fractured rocks. The expressions of water content and relative hydraulic conductivity curves have analytical closed forms. The parameters can be completely determined by the geometry of the fractal model. Every parameter of the proposed formulas of calculating relative hydraulic conductivity of fractured rock has a clear physical meaning. The fractal model can reveal the mechanisms of hydraulic conductivity for fluids flow through unsaturated rocks. The tortuosity fractal dimension D_T which affected other model parameters and should not be neglected while it was often neglected in the past investigation is considered in this paper.

REFERENCES

1. B. Berkowitz and A. Hadad, "Fractal and multifractal measures of natural and synthetic fracture networks," Journal of Geophysical Research B, vol. 102, no. 6, pp. 12205–12218, 1997.
2. R. Moussa, "Is the drainage network a fractal Sierpinskispace," Water Resources Research, vol. 33, pp. 2399–2408, 1997.
3. B. M. Yu, J. C. Cai, and M. Q. Zou, "On the physical properties of apparent two-phase fractal porous media," Vadose Zone Journal, vol. 8, no. 1, pp. 177–186, 2009. · ·
4. B. Q. Xiao, J. T. Fan, and F. Ding, "Prediction of relative permeability of unsaturated porous media based on fractal theory and monte carlo simulation," Energy & Fuels, vol. 26, pp. 6971–6978, 2012.
5. B. Q. Xiao, B. M. Yu, Z. Wang, and L. Chen, "A fractal model for heat transfer of nanofluids by convection in a pool," Physics Letters A, vol. 373, no. 45, pp. 4178–4181, 2009. · ·
6. B. Q. Xiao and B. M. Yu, "A fractal analysis of subcooled flow boiling heat transfer," International Journal of Multiphase Flow, vol. 33, no. 10, pp. 1126–1139, 2007. · ·

7. P. G. Obuko and K. Aki, "Fractal geometry in the San Andreas fault system," Journal of Geophysical Research, vol. 92, pp. 345–355, 1987.
8. J. C. Cai, B. M. Yu, M. Q. Zou, and M. F. Mei, "Fractal analysis of invasion depth of extraneous fluids in porous media," Chemical Engineering Science, vol. 65, no. 18, pp. 5178–5186, 2010. · ·
9. J. C. Cai and B. M. Yu, "Prediction of maximum pore size of porous media based on fractal geometry,"Fractals, vol. 18, no. 4, pp. 417–423, 2010. · ·
10. C. A. Barton and M. D. Zoback, "Self-similar distribution and properties of macroscopic fractures at depth in crystalline rock in the Cajon Pass scientific drill hole," Journal of Geophysical Research, vol. 97, no. 4, pp. 5181–5200, 1992.
11. S. W. Tyler and S. W. Wheatcraft, "Fractal processes in soil water retention," Water Resources Research, vol. 26, no. 5, pp. 1047–1054, 1990. · ·
12. J. S. Wu and B. M. Yu, "A fractal resistance model for flow through porous media," International Journal of Heat and Mass Transfer, pp. 3925–3932, 2007.
13. R. H. Brooks and A. T. Corey, Hydraulic Properties of Porous Media. Hydrology Paper no. 3, lorado State University, Fort Collins, CO, 1964.
14. N. T. Burdine, "Relative permeability from pore size distribution data," Transactions. American Institute of Mining, Metallurgical and Petroleum Engineers, vol. 198, pp. 71–78, 1953.
15. B. Q. Xiao, "Prediction of heat transfer of nanofluid on critical heat flux based on fractal geometry,"Chinese Physics B, vol. 22, no. 1, Article ID 014402, 2013. ·
16. B. Q. Xiao, "A new analytical model for heat transfer in pool boiling," Modern Physics Letters B, vol. 24, no. 12, pp. 1229–1236, 2010. · ·
17. T. M. Nordberg and S. T. Thorolfsson, "Low impact development and bioretention areas in cold climates," in Proceedings of the World Water and Environmental Resources Congress: Critical Transitions in Water and Environmental Resources Management, pp. 3409–3418, Salt Lake City, Utah, USA, July 2004.

18. M. B. Green, J. R. Martin, and P. Griffin, "Treatment of combined sewer overflows at small wastewater treatment works by constructed reed beds," Water Science and Technology, vol. 40, no. 3, pp. 357–364, 1999. · ·

19. D. P. Solomatine and A. Ostfeld, "Data-driven modelling: some past experiences and new approaches,"Journal of Hydroinformatics, vol. 10, no. 1, pp. 3–22, 2008. · ·

20. A. J. Katz and A. H. Thompson, "Fractal sandstone pores: implications for conductivity and formation,"Physical Review Letters, vol. 54, pp. 1325–1328, 1985.

21. W. F. Hunt and A. R. Jarrett, "Evaluating bioretention areas from two field sites in North Carolina," inProceedings of the World Water and Environmental Resources Congress: Critical Transitions in Water and Environmental Resources Management, pp. 797–806, Salt Lake City, Utah, USA, July 2004.

22. B. M. Yu and J. H. Li, "Some fractal characters of porous media," Fractals, vol. 9, no. 3, pp. 365–372, 2001. · ·

23. B. M. Yu and H. M. Li, "A geometry model for tortuosity of flow path in porous media," Chinese Physics Letters, vol. 21, no. 8, pp. 1569–1571, 2004.

Chapter 3

Simulation of Hydraulic Fracture in Unsaturated Soils with High Degree of Saturation

Tielin Chen, Liangyi Zhang, and Dingli Zhang

School of Civil Engineering, Beijing Jiaotong University, Beijing 100044, China

ABSTRACT

A numerical simulation approach of hydraulic fracture process, considering the couplings of the stress distribution, the fluid flow of the water-air mixture, the compression and dissolution of air, and the element damage evolution, has been developed to investigate the mechanisms of crack initiation and propagation in porous media during hydraulic fracturing. The concept of homogenized pore fluid has been adopted to represent the water air mixture. A large number of numerical analysis on hydraulic fracturing in clay with incipient injection slot have been carried out to study the mechanism of hydraulic fracturing in unsaturated soil with the characteristic of critical model I type of crack loading using stress intensity factor . The results provide a numerical picture depicting the mechanisms of crack initiation and propagation during hydraulic fracturing. The numerical results are in good agreement with the experimental results, which confirms the adequacy and the power of the numerical approach.

INTRODUCTION

Hydraulic fracturing may be defined as the process of creating a fracture or fracture system in a porous medium by injecting a fluid under pressure through a well bore in order to overcome native stresses. Hydraulic fracturing has been used for more than 50 years to enhance the yield of wells recovering oil at great depth. In the recent past, hydraulic fracturing methods have been developed for creating fractures in soils and making existing fractures larger to enhance the mass transfer of contaminants. The fractures created increase the effective permeability and change paths of fluid flow, thus making in situ remediation more effective and economical. Hydraulic fracturing appears to have useful environmental, agricultural and geotechnical applications.

Hydraulic fracturing techniques are equally applicable to both saturated soils and unsaturated soils to improve the flow of water and air, respectively. Earth's land is mostly located in the unsaturated vadose zone above the water table, so the hydraulic fracturing of unsaturated soil has important significance. An unsaturated soil is commonly referred to as a 3-phase system composed of solids (soil particles), water, and air. The air-water interface (i.e., the contractile skin) warrants inclusion as an additional phase due to its unique and specific properties. The contractile skin interacts with the soil particles in an independent manner and can significantly change the mechanical behavior of an unsaturated soil. The hydraulic fracturing in unsaturated soil is coupling process of stress distribution, water flow, air flow, water-air interact, and damage evolution, which is arguably one of the most challenging problems.

Especially, the propagation processes of hydraulic fractures enveloped in soil layers are difficult to directly observe, and the details of fracture growth inside soil have generally been inferred from indirect measurements. During the past few years, with the rapid development of computing power, numerical tools have become a good option for gaining some insight into the fracturing process. The main numerical methods coupled with fracture models have been used to simulate fracturing process by many researchers, such as the finite element method (FEM) [1], the mesh-free Galerkin method (EFGM) [2], the extended finite element method (XFEM) [3], the boundary element method (BEM) [4], the discrete element method (DEM) [5], and the displacement discontinuity method (DDM) [6]. The former four are based on the continuum mechanics, the latter two are based on the noncontinuum mechanics.

Hydraulic fracturing is a more complicated process to simulate, as it involves the coupling of four processes [7,8]: (i) the mechanical deformation induced by the fluid pressure on the fracture surfaces; (ii) the flow of fluid within the fracture; (iii) the fracture propagation; and (iv) the coupling of seepage and stress in the computational domain around the fracture. The simulation of hydraulic fracturing in rock and soil mass, arguably one of the most challenging computational problems in geoengineering, has been the subject of numerous investigations since the pioneering work of Khristianovic and Zheltov [9]. Many numerical methods have been developed for the simulation of hydraulic fracturing [10–14]. Among these methods, FEM method is the most robust one in comparison with BEM, DEM, and DDM which cannot efficiently solve the elasticity-plasticity equation relating the fluid pressure to the fracture opening [15].

Two approaches, namely, the discrete crack model and the smeared crack model were introduced to FEM by Ngo and Scordelis [16] and Rashid [17], respectively. The discrete crack model is aimed at simulating the initiation and propagation of dominant cracks. In contrast, the smeared crack model is based on the idea that many small cracks nucleate which only in a later stage of the loading process link up to form one or more dominant cracks.

The discrete crack approach in its original form has several disadvantages [18]. In the approach, the cracks are forced to propagate along the element boundaries, so that a mesh bias is introduced.

Automatic remeshing with sophisticated computer codes allows the mesh bias to be reduced, nevertheless, a computational difficulty, namely, the continuous change in topology, is inherent in the discrete crack approach and is to a certain extent even aggravated by remeshing procedures.

Since each individual crack is not numerically resolved, the smeared crack model captures the deterioration process through a constitutive relation, thus smearing out the cracks over the continuum. It describes a cracked solid by an equivalent anisotropic continuum with degraded material properties in the direction that is normal to the crack orientation and no remeshing is needed. Although the discrete crack approach is sometimes preferred for being based on a mature theory, the smeared crack analysis seems to be more often applicable for usual FEM analyses from engineering practice.

However, many assumptions and simplifications are made in existing numerical studies; for example, fissures need to be placed inside the analyzed domain prior to analysis [19, 20] or the hardening of grout in soil cannot be described [21]. A numerical model for coupled analysis of flow, stress, and damage was proposed to overcome the shortcoming of the placing of preset of fractures [22]. The approach is similar to smeared crack method. The simulated domain is discretized into large numbers of tiny elements to take into account the small local variations of the material properties. At each loading step, the stress and strain in the elements are calculated and examined against the predefined soil strength. Those elements with the stresses above the material's strength were considered to be isotropically damaged. The material properties of the damaged element are reduced and the width of the facture in the damaged element can be calculated. This approach can effectively simulate hydraulic fracturing [22–24] in rocks. In this paper, the extension of the approach for rocks to unsaturated soil is attempted so that hydraulic fracturing in unsaturated soil can be numerically simulated.

In this paper, the Biot's consolidation theory is used as a framework that governs the interactions between the soil skeleton and the pore fluid. The concept of homogenized pore fluid is used to describe water-air mixture fluid. The nonlinear constitutive relationship developed by Duncun and Chang is used to describe the soil deformation for its simplicity. The finite element method (FEM) [25] is employed to discretize the Biot equation. The FEM is chosen because it can readily obtain the solutions of boundary value problems on nonconformal domains.

THE CONCEPT OF AN HOMOGENIZED PORE FLUID

The effective stress equation for saturated soil has been shown to be theoretically valid and practically useful for saturated soils.

$$\sigma = \sigma' + u_w \tag{1}$$

For unsaturated soil, Bishop (1959) [26] suggested the following rational equation:

$$\sigma' = (\sigma - u_a) + \chi (u_a - u_w), \tag{2}$$

in which σ = effective stress, σ = total stress, u_a = pore air pressure, u_w = pore water pressure, and χ = factor whose value ranges from 0 to 1 as degree of saturation varies, and is evaluated experimentally.

However, Sparks [27] has shown that χ can assume values greater than unity, and that it can also be negative. Coleman [28], Blight [29], and Matyas and Radharkrishna [30] have shown that the value of χ applicable to volume change behaviour is different from the value of χ applicable to the strength behaviour of the same soil at the same degree of saturation. This fact indicates that (2) is not fundamentally valid and is not likely to be applicable except under limited ranges of conditions and for limited purposes. Sparks [27] employed a model composed of uniform spheres to develop the following effective stress equation for unsaturated soils:

$$\sigma' = \sigma - \xi_1 u_a - \xi_2 u_w + \xi_3 T_c, \tag{3}$$

in which ξ_1, ξ_2, ξ_3 = parameters, whose values depend on the degree of saturation, and T_c = surface tension of water.

Both of the preceding effective stress equations involve difficulties for practical application. The concept of a "homogenized pore fluid" presented by [31] offers a simple alternative to these equations which is applicable for a limited but useful range of conditions.

The pores of an unsaturated soil are filled partly with gas (air) and partly with liquid (water), and the air phase might be present in an unsaturated soil either in a continuous or an occluded (bubbles) form. An important and frequently encountered special case is that in which the degree of saturation is sufficiently high so that the air bubbles are occluded. This condition prevails at degrees of saturation greater than about 0.85, which is common in many practical cases.

The concept of a "homogenized pore fluid" is that this mixture of gas and liquid can be considered to be an equivalent homogeneous pore fluid which completely fills the pores of the soil. As the proportions of air and water vary (due to flow of water and air, or due to compression of air and dissolution of air), the mechanical properties of the homogenized pore fluid vary. The compressibility of the pore fluid and changes in the degree of saturation due to changes in pore fluid pressure may be calculated using Boyle's Law and Henry's Law [32]. The equation of effective stress may be expressed as

$$\sigma' = \sigma - u_m, \tag{4}$$

in which u_m = effective average pore pressure in the homogenized pore fluid.

Sparks [27] and Barden [33] have found that when the air bubbles are occluded, the pore air pressures and the pore water pressures are very nearly equal. There will inevitably be some small differences between these two pressures as a result of surface tension effects, but these small differences become negligible as the magnitude of the pore pressure increases. Sparks has indicated that for occluded bubbles, the assumption that $u_m = u_w$ introduces very small error. Thus, for this condition, the effective stress equation may be written as follows:

$$\sigma' = \sigma - u_w, \tag{5}$$

in which u_w = pore water pressure. Equation (5) has the same form with (1), but the fluid in (5) is water-air mixture fluid which can be compressed, while (1)'s is incompressible water. Equation (5) is used throughout the following developments, and the results are thus applicable to soils containing water and occluded air bubbles.

THE COMPRESSIBILITY AND PERMEABILITY OF PORE FLUID

The compressibility (β_m) of pore fluid containing air and water is defined as

$$\beta_m = -\frac{d\varepsilon_v}{du_w}, \tag{6}$$

in which $d\varepsilon_V$ = volumetric strain in pore fluid due to change (du_w) in the fluid pressure. By employing Boyle's Law and Henry's Law, the compressibility of an air-water mixture can be expressed in a form suggested by [34]

$$\beta_m = s_r \cdot \beta_w + B_{aw} \cdot \frac{(1 - s_r)}{u_a} + B_{aw} \cdot s_r \cdot \frac{h}{u_a}. \tag{7}$$

The first term in the equation accounts for the compressibility of the water; the second term accounts for Boyle's Law being applied to the free air; and the third term accounts for the air driven into solution in accordance with Henry's Law. The B_{aw} pore pressure parameter is equal to 1 as the air bubbles become occluded, where β_m = compressibility of the air-water mixture, h = volumetric coefficient of air solubility (at 20°C,), β_w = compressibility of the water, which can be ignored because it is a slight amount, and = degree of saturation.

With the increase in pore pressure pore air dissolved in the water phase, the saturation variation can be calculated as [35]

$$s_r = s_{r0} \frac{u_w + p_a}{p_a + (1-h)\, s_{r0} u_w}, \tag{8}$$

where s_{r0} = initial degree of saturation and p_a = atmospheric pressure.

The permeability of unsaturated clays ($k\ w$) depends on the degree of saturation. The relationship between the permeability of saturated soil (k_w) and its degree of saturation can often be approximated using the following relationship [36, 37]:

$$k_w' = k_w \left(\frac{\left(s_r - s_{rf}\right)}{\left(1 - s_{rf}\right)} \right)^{\lambda}, \tag{9}$$

where s_{rf} is the threshold degree of saturation at which the water in the pores begins to flow freely. For a degree of saturation less than s_{rf}, the water forms a membrane around the particles and does not flow.

FIELD EQUATION GOVERNING THE CONSOLIDATION PHENOMENON

Chang and Duncan [31] present an extension of Biot's theory which makes possible finite element analyses of the consolidation of inelastic, unsaturated soil masses. The basic equations governing consolidation may be stated as follows.

- Equilibrium equation. Only changes in stresses and body forces are considered. The geostatic body forces are considered as initial stresses. Consider the following:

$$\sigma_{ij,j} + \rho F_i = 0, \tag{10}$$

where σ_{ij} is the total stress tensor, is the body force per unit of mass, and ρ is the density of the mass.

- Total stresses are the sum of the effective stress $\sigma'_{ij}\sigma ij$ and the pore pressure u. Consider the following:

$$\sigma'_{ij} = \sigma_{ij} - u\delta_{ij}. \tag{11}$$

- Continuity equation. For a compressible pore fluid, the volume decrease of the soil is the sum of the water expelled plus the decrease in volume of the fluid within. The continuity equation can therefore be written as

$$-v_{i,i} + \dot{w}_{i,i} - \beta_m \dot{u} = 0, \tag{12}$$

where v_i is the superficial velocity vector, w_i is the displacement vector, and b_m is the compressibility of the pore fluid. "·" indicates partial differentiation with respect to time.

- The effective stress-strain relationship may be expressed in a general form as follows:

$$\sigma'_{ij} = C_{ijkl}\varepsilon_{kl}, \tag{13}$$

where e_{kl} is the strain tensor, which is related to the displacements through

$$\varepsilon_{kl} = \frac{1}{2}\left(w_{k,l} + w_{l,k}\right), \tag{14}$$

where C_{ijkl} is the tensor relating stress and strain.

- Darcy's Law. It is assumed that the flow of water is governed by Darcy's Law as follows:

$$v_i = k_{ij}\left(u_{,j} + \rho_f F_j\right), \tag{15}$$

where K_{ij} is the permeability tensor, F_j is the body force per unit of fluid, and r_f is the density of the fluid.

To define the problem, both displacement and flow boundary conditions must be specified. For the displacement boundary conditions, it is assumed that part of the boundary surface S_T is subjected to known applied tractions, $\overline{T}_i$, while the remainder of the surface is subjected to specified displacements, $\overline{W}_i$. For the flow boundary condition, it is assumed that part of the boundary surface S_q is subjected to a specified flow velocity, $\overline{q}$, while the remainder of the surface S_u is subjected to known pore pressures, $\overline{u}$. The boundary conditions may be written in the following form:

$$\begin{aligned} w_i &= \overline{w}_i \quad \text{on } S_D \text{ for } t \geq 0, \\ \sigma_{ij}n_j &= \overline{T}_i \quad \text{on } S_T \text{ for } t \geq 0, \\ u &= \overline{u} \quad \text{on } S_u \text{ for } t > 0, \\ v_i n_i &= \overline{q} \quad \text{on } S_q \text{ for } t > 0, \end{aligned} \tag{16}$$

where n_i is the normal vector to the boundary surface and (−) indicates a prescribed quantity.

CONSTITUTIVE EQUATION

The nonlinear elastic model proposed by Duncan and Chang is used as the constitutive relationship for solids. The main features of the model are summarized here. The tangential Young's modulus can be expressed as

$$E_t = Kp_a\left(\frac{\sigma_3}{p_a}\right)^n\left[1-\frac{R_f(1-\sin\varphi)(\sigma_1-\sigma_3)}{2c\cos\varphi+2\sigma_3\sin\varphi}\right]^2. \tag{17}$$

The elastic Young's modulus for unloading and reloading is

$$E_{ur} = K_{ur}p_a\left(\frac{\sigma_3}{p_a}\right)^n. \tag{18}$$

The tangential volumetric modulus is described by

$$B = K_b p_a\left(\frac{\sigma_3}{p_a}\right)^n, \tag{19}$$

where c is cohesion, δ is internal friction angle, p_a is atmospheric pressure, K, K_b, K_{ur} are initial modulus values, m and are parameters, and R_f is a failure ratio that is defined as

$$R_f = \frac{(\sigma_1-\sigma_3)_f}{(\sigma_1-\sigma_3)_{ult}}, \tag{20}$$

where $(s_1 - s_3)_f$ is the principal stress difference when the soil is assumed to be in failure and $(s_1 - s_3)_{ult}$ is the asymptotic value of the principal stress difference.

Eight parameters of the constitutive model c, φ, K, K_b, K_{ur}, m, n, and R_f can be determined by conventional triaxial tests.

MECHANICS OF FRACTURE INITIATION AND PROPAGATION

The tensile and shear failures are both considered here. An element is considered to be fractured in the tension mode if its minor effective principal stress exceeds the tensile strength of the element as follows:

$$\sigma_3' \leq -\sigma_t', \tag{21}$$

where σ_3' is the minor effective principal stress and is the tensile failure strength of the element. Compression is assumed to be positive.

An element is considered to be fractured in the shear mode if the shear stress satisfied the Mohr-Coulomb failure criterion as follows:

$$F = \left(c' + \sigma' \tan \phi'\right) - \tau, \tag{22}$$

where τ is the shear stress, σ' the effective normal stress, C′the effective stress cohesion intercept, and the effective stress angle of friction or shearing resistance.

Once an element is judged to be fractured, that is, mesh element damaged, many smeared cracks occur in this element and the stiffness of the soil is reduced.

The stiffness of the fracture can be described by [38]

$$K' = Ke^{-b/\xi}, \tag{23}$$

where K is the initial modulus of the soil just prior to the occurrence of cracks, b is the total width of smeared cracks in an element (m), and ξ is a scaling parameter indicating the amount of deformation needed to change the stiffness by a factor of K_b. K_{ur} also follow the formula. The total width of smeared cracks is the difference between the widths of the element after and before cracking. The value of ξ is related to the material types and is 10^{-4} in this investigation.

The permeability is not a constant and is a function of stresses because the fracture aperture is most likely to change as the stress conditions vary. The effect of stresses on permeability can be expressed as [22]

$$k(\sigma, u_w) = \omega_1 k'_w \exp\left(-\omega_3\left(\frac{\sigma_{ii}}{3} - \omega_2 u_w\right)\right), \tag{24}$$

where ω_1 is the damage factor of permeability reflecting the increase in permeability induced by damage of soil, ω_2 and ω_3 are coupling parameters reflecting the influence of stress on the coefficient of permeability.

THE SIMULATION OF K C OF UNSATURATED SOIL DURING HYDRAULIC FRACTURE

The critical stress at the onset of propagation depended on the length of the initial slot. The strength was characterized by the stress intensity factor K_i [39, 40] as follows:

$$K_I = \sigma_c \sqrt{\pi a} f(b), \tag{25}$$

where σ_c is far-field stress, a is the half-length of the fracture, and f(b) is some function of the geometry of the fracture and enveloping material.

A basic assumption is that propagation under dilation, or mode I, loading occurs when the stress intensity equals a critical value K_{Ic}. The magnitude of K_{Ic} is commonly regarded as a material property characterizing the resistance of a material to mode I fracturing; here it is called fracture toughness.

Fracture toughness has been applied to prediction of the onset of hydraulic fracturing of soil [41–43], but the validity of that application needs to be examined further.

The stress intensity factor of soil can be defined as follows:

$$K_I = P_d \sqrt{\pi a} f\,(b)\,. \tag{26}$$

Driving pressure p_d is the difference between the fluid pressure p within a fracture and the confining stress σ_n acting normal to the fracture as follows:

$$P_d = p - \sigma_n. \tag{27}$$

The function f(b) in (26) is equal to unity if the fracture is straight, loaded uniformly, and embedded in a homogeneous, isotropic, linearly-elastic material, and it can be taken as a correction factor for practical constraints such as finite sample size or slots intersected by a hole. Singh [37] have given expressions for f(b) for a straight slot containing an axial hole and for a straight slot embedded in a rectangular sample of finite size. These conditions represent corrections of <5% for the geometries and dimensions used in this work, so f(b) will be set to unity. So, fracture toughness will be determined using the critical driving pressure p_{df} and the half-length of the initial slot a_i as

$$K_{Ic} = P_{df} \sqrt{\pi a_i}. \tag{28}$$

The propagation of hydraulic fractures in soil was studied in the laboratory by glycerine injected at a constant rate into rectangular samples of silty clay confined in a triaxial cell [41–43]. In order to examine the validity of the application, a large number of numerical analyses on hydraulic fracturing in clay with incipient injection slot have been carried out to investigate the relation between and . And the results are compared with experimental results [41–43].

A simulation domain 0.1 m × 0.39 m with incipient injection slot is the same with the test [41–43]. The soil mass within the simulated area is under the plane strain boundary condition, as shown in Figure 1.

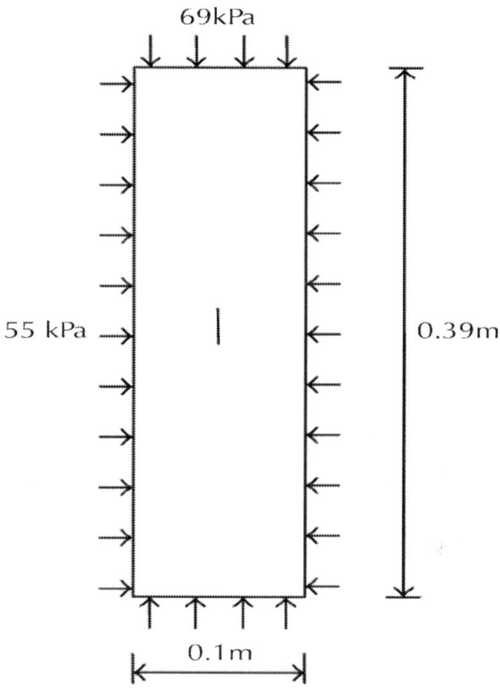

Figure 1: The simulation model.

The relationship between initial modulus values and saturation can be obtained from the experimental results [41] as follows:

$$K = K_0 - 5K_0\left(s_r - 0.85\right),$$

$$K_b = K_{b0} - 5K_{b0}\left(s_r - 0.85\right),$$

$$K_{ur} = K_{ur0} - 5K_{ur0}\left(s_r - 0.85\right). \tag{29}$$

The input parameters for these simulations are tabulated in Table 1.

Table 1: Input material properties parameters for numerical models

Index	Value
$K0$	120
$Kb0$	204
$Kur0$	240
m	0.5
n	0.6
Rf	0.95
c	15 kPa
φ	25°
ξ	10–4
λ	3
Srf	0.3
Kw	cm/s
α	3
$\omega1$	20
$\omega2$	0.1
$\omega3$	0.05

Six cases of the initial saturation from 0.85 to 0.95 at interval 0.02, 12 cases of length of the starter slot from 0.008 m to 0.03 m at interval 0.002 m, 2 cases of consolidation durations of 4 days and 14 days are considered. A flow rate of 0.033 cm^3/s which was the same with the test was used for all the simulations described below.

The numerical analysis can capture details of the k_{Ic} at various saturation in the range >85%. The result is shown in Figures 2 and 3. The laboratory records [42] and the simulations in Figures 2 and 3 show a remarkable similarity. The simulations can accommodate k_{Ic} changes caused by the saturation and the length of start slot.

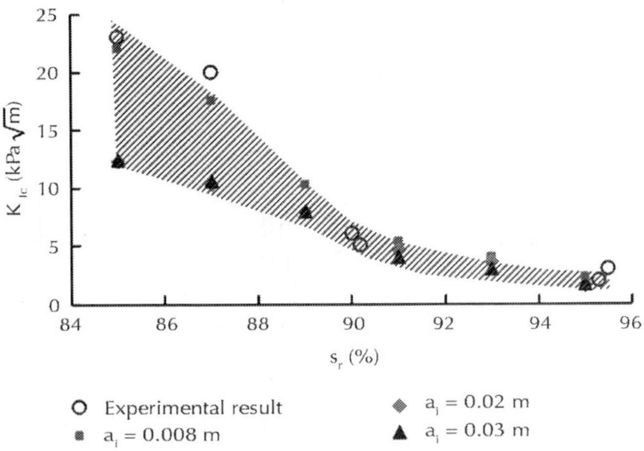

Figure 2: k_{Ic} as a function of S_r.

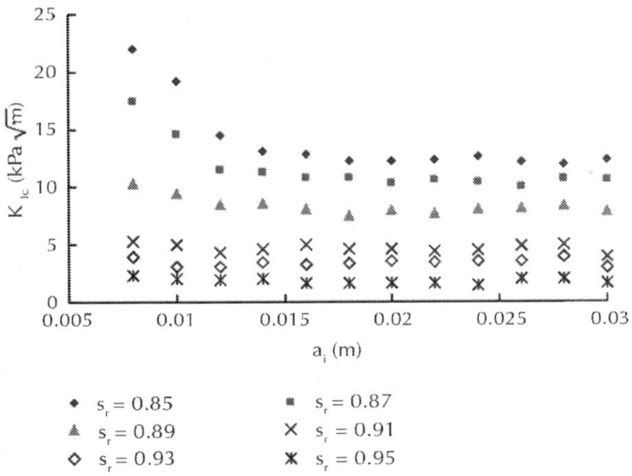

Figure 3: k_{ic} as a function of a_i.

Both saturation and duration of consolidation had a marked effect on k_{Ic}. Among samples consolidated for 14 days, for a_i= 0.008 m, k_{Ic} is greatest (22 kPa√m) at s_r= 0.85.. The fracture toughness decreases abruptly to 2.33 as s_r increases from 0.85 to 0.95. For a_i= 0.03 m, the fracture toughness decreases abruptly from 12.4 kPa√m to 1.68 kPa√m The sharp change in k_{Ic} occurs in the range $0.85 \leq s_r < 0.91$. A slight decrease in k_{Ic} occurs with further increase in s_r, although the decrease is small and k_{Ic} is almost independent of in the range 0.91–0.95.

Consolidation appears to toughen the clay, according to simulation results of samples consolidated for 4 days and 14 days under a static pressure of 69 kPa. The k_{Ic} of the samples consolidated for 14 days is greater by 6.1 kPa√m than the k_{Ic} of 4 days consolidated samples at similar s_r and a_i=0.008 m. With the increase of saturation, the difference of k_{Ic} is reduced, which means that the tough effect of consolidation is weakened.

The forms of the experimental and simulation curves in Figures 2 and 4 are similar; they all show k_{Ic} decreasing markedly over a few percent of s_r.

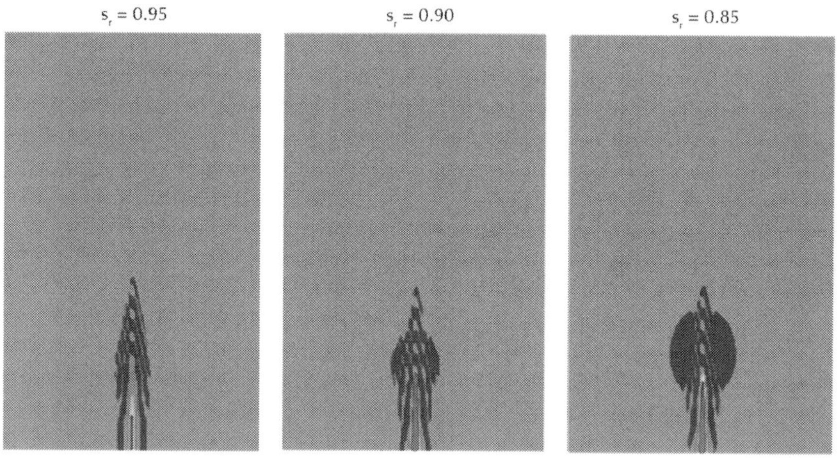

Figure 4: Negative pore pressure zone (dark blue color) at different saturations arranged in columns.

Fracture toughness appears to be independent of the length of the starter slot in the range $s_{r > 0.91}$ and to be dependent of the length of the starter slot in the range 0.85 $s_{r > 0.91}$ under the conditions used here. Increase of the length of the starter slot can decrease this dependence.

As we all know, negative pore pressure can affect the strength of the soil. The area of negative pore pressure zone (NPZ) may explain the dependence between fracture toughness and saturation and length of starter slot. Figure 4 shows the NPZ before facture tip. Only half of the simulated domain is displayed because of symmetry. When the saturation is higher (0.95), the area of NPZ and the length of the starter slot are almost independent of each other (Figure 4). When the saturation is lower (0.85), the length of the starter slot has a greater impact on the area of NPZ (Figure 4).

Fracture toughness has been introduced as a parameter that could possibly be used to indicate the onset of propagation. If it is a material property, fracture toughness must be independent of slot length. It can depend on saturation, however, although the dependence is expected to be similar to other properties that also depend on saturation.

From the simulation results, the k_{Ic} depends on the length of the initial slot and the saturation of the soil. It is apparent from Figures 2 and 3 that the value of k_{Ic} varies over more than a factor of ten, so that some methods of anticipating k_{Ic} will be required in order for this to be a useful predictor of the onset of propagation.

THE SIMULATION OF HYDRAULIC FRACTURE IN HETEROGENEITY UNSATURATED SOIL

Actual soil is often heterogeneous, in order to demonstrate the effectiveness of this approach to simulate hydraulic fracture in heterogeneous unsaturated soil, three simulations are carried out. A simulation domain 0.1 m × 0.1 m with initial injection tiny holes in the center is considered here. All boundaries are fixed displacement, undrained boundary. The distribution of heterogeneity remains the same and the saturation varies from one sample to another.

To take the heterogeneity into account, different material properties for different elements are assigned randomly following the Weibull distribution as follows:

$$\psi = \frac{\vartheta}{\eta_0}\left(\frac{\eta}{\eta_0}\right)^{\vartheta-1} \exp\left[-\left(\frac{\eta}{\eta_0}\right)^{\vartheta}\right], \tag{30}$$

where η is the variable representing a certain material property, η_0 is the mean value of the corresponding material property, and ϑ is the homogeneity index. ϑ is a parameter defined by the shape of the distribution function that describes the degree of material heterogeneity. A larger ϑ implies a more homogeneous material and vice versa. Therefore, the parameter ϑ is called the homogeneity index. The higher the value of the homogeneity index , the greater the number of elements with properties close to . η_0 In the study, K_0, K_{b0}, r_0, c, φ are accorded to the Weibull distribution by $\partial = 3$.

Figure 5 shows the simulated fracture propagation process. Each column is a simulation result at a saturation. Saturation has a significant impact on fracture propagation and distribution. When saturation is high (s_r= 0.95), numerous cracks gathered near the initial hole and cross-cuttings occur between the cracks. The lower the saturation, the farther away the cracks spead from the initial hole and the fewer the number of cracks there is.

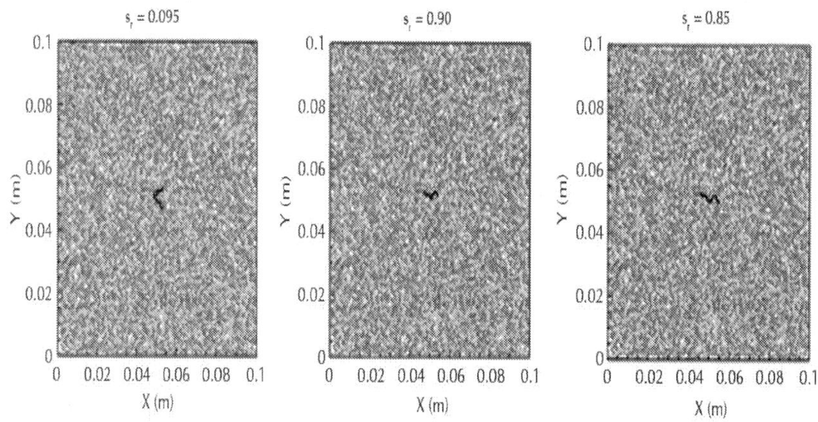

Figure 5: The simulated fracture propagation process at different saturations arranged in columns.

Hydraulic fracturing appears to have useful environmental and geotechnical applications, but the details of fracture morphology and propagation in soil are poorly known because the propagation processes of hydraulic fractures enveloped in soil layers are difficult to directly observe. The simulation results may provide some insight.

CONCLUSIONS

A numerical simulation approach of hydraulic fracture process of unsaturated soil with high saturation has been presented. The approach can consider the coupling of stress distribution, water-air mixture fluid flow, compression and dissolution of air, and element damage evolution. A key to the approach is the concept of homogenized pore fluid which is that the mixture of gas and liquid can be considered to be an equivalent homogeneous pore fluid which completely fills the pores of the soil. As the proportions of air and water vary (due to flow of water and air, or due to compression of air and dissolution of air), the mechanical properties of the homogenized pore fluid vary correspondingly.

A large number of numerical analyses on hydraulic fracturing in unsaturated soil with incipient injection slot have been carried out to study the characteristics of critical stress intensity factor K_{Ic} of model I type of crack loading. The numerical results were shown to be in good agreement with the experimental results. The propagation of hydraulic fracturing in unsaturated soil has been simulated also. The simulation results have indicated that saturation is a fundamental constraint on the hydraulic fracturing in clay.

Although this study is a preliminary attempt and needs further development, the simulation results may provide some insight of the crack initiation and propagation mechanisms during hydrofracturing of unsaturated soil with high saturation.

ACKNOWLEDGMENTS

The authors gratefully acknowledge the financial support of the National Basic Research Program of China under Grant 2010CB732100 and the National Natural Science Foundation of China under Grant 51278041.

REFERENCES

1. R. S. Barsoum, "Further application of quadratic isoparametric finite elements to linear fracture mechanics of plate bending and

general shells," International Journal of Fracture, vol. 11, no. 1, pp. 167–169, 1975.

2. B. N. Rao and S. Rahman, "Efficient meshless method for fracture analysis of cracks," Computational Mechanics, vol. 26, no. 4, pp. 398–408, 2000.
3. B. Lecampion, "An extended finite element method for hydraulic fracture problems," Communications in Numerical Methods in Engineering, vol. 25, no. 2, pp. 121–133, 2009.
4. A. Frangi, G. Novati, R. Springhetti, and M. Rovizzi, "3D fracture analysis by the symmetric Galerkin BEM," Computational Mechanics, vol. 28, no. 3-4, pp. 220–232, 2002.
5. K.-B. Min and L. Jing, "Numerical determination of the equivalent elastic compliance tensor for fractured rock masses using the distinct element method," International Journal of Rock Mechanics and Mining Sciences, vol. 40, no. 6, pp. 795–816, 2003.
6. X. C. Tan, S. Q. Kou, and P.-A. Lindqvist, "Application of the DDM and fracture mechanics model on the simulation of rock breakage by mechanical tools," Engineering Geology, vol. 49, no. 3-4, pp. 277–284, 1998.
7. J. Adachi, E. Siebrits, A. Peirce, and J. Desroches, "Computer simulation of hydraulic fractures,"International Journal of Rock Mechanics and Mining Sciences, vol. 44, no. 5, pp. 739–757, 2007.
8. A. P. Bunger, E. Detournay, and D. I. Garagash, "Toughness-dominated hydraulic fracture with leak-off," International Journal of Fracture, vol. 134, no. 2, pp. 175–190, 2005.
9. S. A. Khristianovic and Y. P. Zheltov, "Formation of vertical fractures by means of highly viscous fluids," in Proceedings of the 4th World petroleum congress, vol. 2, pp. 579–586, Rome, Italy, 1955.
10. A. P. Peirce and E. Siebrits, "A dual mesh multigrid preconditioner for the efficient solution of hydrautically driven fracture problems," International Journal for Numerical Methods in Engineering, vol. 63, no. 13, pp. 1797–1823, 2005.
11. B. Lecampion and E. Detournay, "An implicit algorithm for the propagation of a hydraulic fracture with a fluid lag," Computer

Methods in Applied Mechanics and Engineering, vol. 196, no. 49–52, pp. 4863–4880, 2007.

12. L. Jing, Y. Ma, and Z. Fang, "Modeling of fluid flow and solid deformation for fractured rocks with discontinuous deformation analysis (DDA) method," International Journal of Rock Mechanics and Mining Sciences, vol. 38, no. 3, pp. 343–355, 2001.
13. K. I.-I. Eshiet, Modelling of hydraulic fracturing and its engineering application [Ph.D. thesis], University of Leeds, 2012.
14. Z. P. Bazant and H. Ohtsubo, "Geothermal heat extraction by water circulation through a large crack in dry hot rock mass," International Journal for Numerical and Analytical Methods in Geomechanics, vol. 2, no. 4, pp. 317–327, 1978.
15. M. B. Dusseault and J. McLennan, "Massive multi-stage hydraulic fracturing: where are we?" 2011, ARMA, http://www.armarocks.org/documents/newsletters/2011_01_02_winter.pdf.
16. D. Ngo and A. C. Scordelis, "Finite element analysis of reinforced concrete beams," Journal of the American Concrete Institute, vol. 64, pp. 152–163, 1967.
17. Y. R. Rashid, "Analysis of reinforced concrete pressure vessels," Nuclear Engineering and Design, vol. 7, pp. 334–344, 1968.
18. R. de Borst, J. J. C. Remmers, A. Needleman, and M.-A. Abellan, "Discrete vs smeared crack models for concrete fracture: bridging the gap," International Journal for Numerical and Analytical Methods in Geomechanics, vol. 28, no. 7-8, pp. 583–607, 2004.
19. B. Carrier and S. Granet, "Numerical modeling of hydraulic fracture problem in permeable medium using cohesive zone model," Engineering Fracture Mechanics, vol. 79, pp. 312–328, 2012.
20. C. Wisser, C. E. Augarde, and H. J. Burd, "Numerical modelling of compensation grouting above shallow tunnels," International Journal for Numerical and Analytical Methods in Geomechanics, vol. 29, no. 5, pp. 443–471, 2005.
21. M. J. Yang, Z. Q. Yue, P. K. K. Lee, B. Su, and L. G. Tham, "Prediction of grout penetration in fractured rocks by numerical simulation," Canadian Geotechnical Journal, vol. 39, no. 6, pp. 1384–1394, 2002.

22. C. A. Tang, L. G. Tham, P. K. K. Lee, T. H. Yang, and L. C. Li, "Coupled analysis of flow, stress and damage (FSD) in rock failure," International Journal of Rock Mechanics and Mining Sciences, vol. 39, no. 4, pp. 477–489, 2002.
23. S. Y. Wang, L. Sun, A. S. K. Au, T. H. Yang, and C. A. Tang, "2D-numerical analysis of hydraulic fracturing in heterogeneous geo-materials," Construction and Building Materials, vol. 23, no. 6, pp. 2196–2206, 2009.
24. W. Peng, M. Xian-biao, L. Jin-bin, and D. Chun-zhi, "Study of the borehole hydraulic fracturing and the principle of gas seepage in the coal seam," in Proceedings of the 6th International Conference on Mining Science and Technology (ICMST '09), pp. 1561–1573, October 2009.
25. K. J. Bathe, Finite Element Procedures, Prentice-Hall, Englewood Cliffs, NJ, USA, 1996.
26. A. W. Bishop, "The principle of effective stress," Teknisk Ukeblad, Norway, vol. 106, no. 39, pp. 859–863, 1959.
27. A. D. W. Sparks, "Theoretical considerations of stress equations for partly saturated soils," inProceedings of the 3rd Regular Conference for Africa on Soil Mechanics and Foundation Engineering, Salisbury, Southern Rhodesia, 1963.
28. J. D. Coleman, "Stress/strain relations for partly saturated soil," Geotechnique, vol. 12, pp. 348–350, 1962.
29. G. E. Blight, "A study of effective stress for volume change," in Proceedings of the Symposium on Moisture Equilibrium and Moisture Changes in Soils Beneath Covered Areas, pp. 259–269, Butterworths, Australia, 1965.
30. E. L. Matyas and H. Radharkrishna, "Volume change characteristics of partially saturated soils,"Geotechnique, vol. 18, no. 4, pp. 432–448, 1968.
31. C. S. Chang and J. M. Duncan, "Consolidation analysis for partly saturated clay by using an elastic-plastic effective stress-strain model," International Journal for Numerical & Analytical Methods in Geomechanics, vol. 7, no. 1, pp. 39–55, 1983.
32. A. W. Skempton and A. W. Bishop, "Building materials, their elasticity and inelasticity," in Soils, North Holland, Amsterdam, The Netherlands, 1954.

33. L. Barden, "Consolidation of compacted and unsaturated clays," Geotechnique, vol. 15, no. 3, pp. 267–286, 1965.
34. D. G. Fredlund, "Density and compressibility characteristic of air-water mixtures," Canadian Geotechnical Journal, vol. 13, no. 4, pp. 386–396, 1976.
35. J. W. Hilf, "Estimating construction pore pressure in rolled earth dams," in Proceedings of the 2nd International Conference on Soil Mechanics and Foundation Engineering, pp. 234–240, Rotterdam, The Netherlands, 1948.
36. C. S. Chang, Analysis of consolidation of earth and rockfill dams [Ph.D. thesis], University of California, Berkeley, Calif, USA, 1976.
37. R. Singh, "Unsteady and unsaturated flow in soils in two dimensions," Tech. Rep. 54, Department of Civil Engineering, Stanford University, 1965.
38. R. Gothäll and H. Stille, "Fracture-fracture interaction during grouting," Tunnelling and Underground Space Technology, vol. 25, no. 3, pp. 199–204, 2010.
39. G. R. Irwin, "Analysis of stresses and strains near the end of a crack traversing a plate," Journal of Applied Mechanics, vol. 24, pp. 361–364, 1957.
40. H. Tada, P. C. Paris, and G. R. Irwin, The Stress Analysis of Cracks Handbook, Paris Productions, 2nd edition, 1985.
41. L. C. Murdoch, "Hydraulic fracturing of soil during laboratory experiments—1. Methods and observations," Geotechnique, vol. 43, no. 2, pp. 255–265, 1993.
42. L. C. Murdoch, "Hydraulic fracturing of soil during laboratory experiments—2. Propagation,"Geotechnique, vol. 43, no. 2, pp. 267–276, 1993.
43. L. C. Murdoch, "Hydraulic fracturing of soil during laboratory experiments—3. Theoretical analysis,"Geotechnique, vol. 43, no. 2, pp. 277–287, 1993.

Chapter 4

Mechanical Origin of B-value Changes during Stimulation of Deep Geothermal Reservoirs

Arne Stormo, Olivier Lengliné, and Jean Schmittbuhl

EOST, Université de Strasbourg/CNRS, 5 rue René Descartes, Strasbourg 67000, France

ABSTRACT

Background

The distribution of induced-earthquake magnitudes in deep geothermal reservoirs is a classical tool for monitoring reservoirs. It typically shows some important fluctuations through time and space. Despite being a very crude information (i.e., a scalar quantity) of very complex mechanical stress evolution, understanding these variations could still give us insights into the mechanics of the reservoir. Here, we analyze the output of a simple quasi-static physical model of a single fault and propose a new way to describe bursts that could be compared to seismic events.

Methods

Our model is an elastic fiber bundle model describing multiple ruptures along a single major fault of the reservoir. It consists of a set of fibers with various strengths, arranged in a planar $L \times L$ two-dimensional array, linking together two elastic half-spaces. It mimics a fault zone with various asperities. During load, the fibers break quasi-statically according to a stress threshold distribution. Contrary to classical fiber bundle model, here, when a fiber breaks, it redistributes the load on the surviving fibers through long range elastic interactions. Interestingly, the elasticity of the half-spaces which changes the range of the stress distribution, characterizes two distinct regimes. In a stiff regime, we find an effective ductile deformation regime at large scale since the damage distribution is very difuse. Conversely in softer systems, the distance between two consecutive breaking fibers gets smaller and failure of the interface is localized, exhibiting an effective brittle regime.

Results

We analyze two types of burst distributions: a classical one built from the statistics of the broken bonds during each failure step and a new one defines from a waiting time matrix of the fracture front propagation. The first one reproduces several known results for this type of model. The new one evidences the existence of effective creeping advances of the front with statistics that follow a Gutenberg-Richter distribution, in particular, in the ductile regime (stiff systems).

Conclusions

We proposed a new definition of bursts in a fault model, based on the local fracture front velocity. We find that b_v-values for the distribution of the velocity clusters are very consistent with Gutenberg-Richter distribution of induced seismicity. We link the b_v-value fluctuations in the reservoirs to the influence of the velocity threshold level that could be related to recording limitation. b_b-values obtained from the broken bond statistics are hardly comparable to seismic events because of the lack of space contiguousness of broken fibers during bursts. In the light of our results, we discuss the implications of *b*-value changes in geothermal reservoir in terms of fault asperities and normal stress evolution.

BACKGROUND

Earthquakes recorded worldwide from various natural fault environments exhibit a specific magnitude distribution. This distribution is well represented by the Gutenberg-Richter empirical law which states that:

$$\log_{10} N(M) = a - bM, \tag{1}$$

where $N(M)$ is the number of events with a magnitude greater or equal to the magnitude M, and *a*and *b* are the parameters that are inverted from the earthquake magnitude dataset [Gutenberg and Richter (1944])]. The parameter *a* scales with the number of earthquakes in the investigated catalog while the parameter *b* is an exponent which quantifies the overall proportion between larger and smaller magnitude events in the earthquake catalog. The validity of Equation 1 has been observed over a wide range of scales. The scale-free nature of the Gutenberg-Richter relation suggests that laws governing seismic activity at large scales (i.e., fault scales) are the same as at small scales (i.e., laboratory scale) for rock samples [Scholz CH (2002])]. Although the empirical relation of Equation 1 is robust and tested against a large number of dataset of natural earthquakes and acoustic emissions in laboratory experiments, the exponent *b* exhibits fluctuations. While generally close to b=1.0, e.g., [Schorlemmer et al. (2005]) and [Kwiatek et al. (2010]), several variations of the *b* value are also noticed, e.g., [Schorlemmer et al. (2005]),[Enescu and Ito (2001]), and [Amitrano (2003]). These fluctuations attest for a possible physical control on the value of this exponent and its observed fluctuations.

Apart from natural systems and laboratory experiments, the Gutenberg-Richter distribution of earthquakes is also observed for induced seismicity. One example of the later is the case of the stimulation of deep geothermal reservoir. Indeed, for the development of deep geothermal reservoirs (Enhanced Geothermal Systems (EGS)) like at Soultz-sous Forêts, France [Dorbath et al. (2009])] or Basel, Switzerland [Bachmann et al. (2011])], hydraulic injections have been used to stimulate the existing fracture network both in the sediments and in the granite massif in order to enhance the permeability of the reservoir. For instance, during the hydraulic injection of the Soultz-sous-Forêts GPK2 well in 2000, a significant seismic activity was generated, mapping the development of the fault network [Cuenot et al. (2008])]. Up to 30,000 earthquakes were recorded with a magnitude ranging from −0.9 to 2.6. Earthquake magnitude distribution was found to follow the typical Gutenberg-Richter distribution. During the 2000 injection, the *b*-value was measured to be 1.29 for the entire period [Cuenot et al. (2008])]. However, when searching for temporal variations, the *b*-value was observed to evolve from 1.0 to 1.7. High injection rates led to higher*b*-value. Similar variations of the *b*-value during injection were observed in Basel during the 2006 experiment. During the injection, *b*-value was measured to be 1.58 but lowered to 1.15 after shut-in [Bachmann et al. (2011])]. These temporal and spatial fluctuations of the *b*-value suggest to use this information as a tracer of the evolution of the the reservoir (in the case of induced seismicity) or the fault system (in natural environment).

Numerous models have been proposed to explain the variability of the *b*-value. One of the first model is a theoretical study of rock fracture developed by [Scholz (1968]). This model was based on a heterogeneous distribution of material strength in a sample under load. It was found that the*b*-value is a function of the applied stress, and that the frequency of fracture size was scale free. Similar results were obtained by [Wyss (1973]) suggesting that low *b*-values correlate with high stresses. Although these models bring a possible explanation of the observed *b*-value fluctuations, some care should be taken when interpreting stress from the *b*-value alone since it is a scalar quantity that cannot characterize completely the stress tensor. Many other models concerning the connection between earthquakes and micro-fractures have been developed through the years in the language of statistical physics, e.g., [Holliday et al. (2008]). In particular, approaches that connect fiber-bundle models to acoustic emissions during rock fracturing have been done [Turcotte et al. (2003]); [Pradhan et al.

(2010])]. For an overlook of this particular field, we refer to the review by [Rundle et al. (2003]).

Several attempts have also been made in order to model the *b*-value and its fluctuations specifically for induced seismicity in geothermal system, e.g., Hakimhashemi et al. ([2014]).[Baisch et al. (2010]) considered a single fault to model the induced seismicity at Soultz-sous-Forêts. They indeed noticed that hypocenters are aligned along a preexisting planar fault structure which concentrates most of the recorded seismicity. Their quasi-static model includes sliders connected by springs that incorporate elastic stress redistribution. The effect of fluid is simply modeled as a change of pore pressure bringing each path of the model closer to failure in the Mohr-Coulomb sense. Based on this simple model, they reproduced most of the seismicity characteristics: migration of the seismicity, Kaiser effect and magnitude distribution. Variation of the *b*-value was also explained as the result of a dominant portion of smaller cracks close to the injection well, compared to the rock mass away from the well (far-field), e.g., [Zang et al. (2014]). Some more complicated models designed to incorporate more realistic fluid transfer were also proposed. In particular, [Yoon et al. (2014]) proposed a model of aggregated circular particles where both mode I and mode II ruptures are taken into account to compute fluid-induced seismicity in naturally fractured granitic EGS reservoirs. Their modeling was capable of reproducing most of the observed features of the seismicity including the frequency magnitude distribution. Similar results were also obtained by [Gischig and Wiemer (2013]) while using a completely different model. They incorporate a stochastic seed model generating synthetic seismic catalogs reproducing the Basel seismicity data. Other continuous models that include elastic long range interactions together with poro-elastic behavior have also been developed and show self-induced seismicity when shear rupture is dominant over the fluid migration process [Aochi et al. (2014])].

It then appears that many different models from very different physical mechanisms are able to reproduce the frequency magnitude distribution of the seismicity in EGS reservoirs. It therefore questions about how to interpret the change of *b*-values in the light of so different modeling. We decided here to study another aspect: the evolution of damage and its link with seismic events considering only a quasi-static model, where damages are represented on a single fault plane and where the fluid source is not explicitly modeled. Although our approach is oversimplified compared to the rich complexity of induced seismicity of geothermal system, it allows us to test the direct influence of several parameters such as the stiffness of the system. Our model is inspired by the model developed by [Amitrano et al. (1999]) that has shown very promising results in connection to observed seismicity. In this model, load is applied to an initially intact sample. When the stress exceeds the local strength, the sample is damaged. The effect of this damage is represented by a change in local Young's modulus. Thus, increasing the stress in the vicinity of the damage zone increases the probability of damage to the neighboring area. This model has shown to have a brittle to ductile transition as a function of the internal friction angle of the Mohr-Coulomb failure criterion [Amitrano (2003])]. In the ductile regime (i.e., low internal frictional angle), the damage in the fracture process is highly distributed, while the failure in the brittle regime is highly localized for a high internal frictional angle. The transition also affects the *b*-value of the model. Completely ductile failure gives a *b*-value of 0.63, while for brittle fractures, b=1.2. A recently developed bottom-up model of fracture by [Stormo et al. (2013]) has shown to have similar diffuse to localized damage distribution transition. This model can address the competition between elastic stress redistribution and damage development related to geological heterogeneity. Here, we refine and expand the model proposed by Stormo et al. ([2013]) with the attempt to describe fracture development along a pre-existing weak interface. We observe in our model a similar effective ductile to brittle transition at large scale as observed by previous authors. We focus on the analysis of burst distributions. We notably explore the impact of the definition of the event in the results of the frequency magnitude distribution. We indeed introduced a new definition of a burst based on the local velocity of the crack front propagation. We show that this definition provides an interesting perspective for comparison with large scale fluctuations of the *b*-values.

METHODS

Our model is a variation of the fiber bundle model, i.e., a model of a single fracture described as an interfacial failure along a discrete set of connecting fibers. We assume that the fault network in the geothermal reservoir is dominated by a discrete set of pre-existing faults as it was shown by ([Sausse et al. 2010]) at Soultz-sous-Forêts. We even push further the hypothesis and reduce the fault network for seismicity modeling to a single main fault zone as previously proposed by [Baisch et al. (2010]). The response of the main fault is seen as a collection of weak fibers attached to the fault boundaries displaced apart by a distance *D* because of hydraulic pressure increase. Our variation of the fiber bundle model is to deal with the elastic fiber bundle model, first proposed by [Batrouni et al. (2002b]). It is an attempt to include non-local elastic mechanics to the fiber bundle models by replacing the rigid boundaries of the global load sharing configuration by deformable elastic half-spaces. Actually, this configuration of two facing elastic half-spaces is equivalent to a single elastic half-space with equivalent elastic moduli facing a rigid block as discussed in [Batrouni et al. (2002b]). Variability of the fracture aperture could have been introduced by considering a specific morphology of the rigid part [Hansen et al. (2000])], but this is not the case in the present study. However, we consider the presence of a static fluid in the fracture by dealing with normal effective stresses on the fibers but neglect other specific fluid effects like viscous pressure or dynamic effects [Jaeger et al. ([2009])].

The Elastic Fiber-Bundle Model

In our model that follows [Batrouni et al. (2002b]), the fibers are arranged in a planar two-dimensional square array of $L \times L = N$ fibers distanced apart from their nearest neighbors by a distance *a*. Our square array is considered periodic in both in-plane directions. The fibers are attached to two infinite half-spaces, representing the clamps holding the fibers. One of these half-spaces is supposed to be elastic with a Young's modulus *E* and Poisson's ratio . The other one is assumed to be perfectly rigid. This configuration is actually equivalent to two facing elastic half-spaces as discussed in [Batrouni et al. (2002b])]. Figure 1 gives an illustration of the model setup.

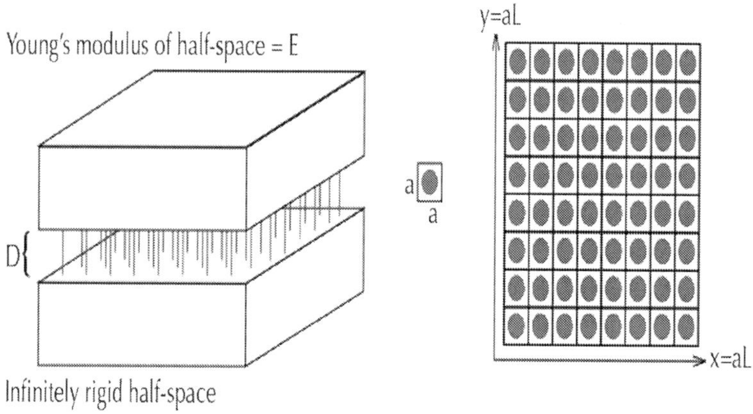

Figure 1: Setup of the elastic fiber bundle model. On the left, a sketch of the fiber bundle. The upper half-space deflects in response to the forces applied by the fibers (red), while the infinitely rigid lower half-space is displaced downward of a distance *D*. To the right, a schematic top view of the entire fiber bundle where each red circle represents a fiber at the center of a mesh of size *a*×*a*.

Each fiber is modeled as a Hookean spring, and thus exercises an opposite force, *f*, on a square $a^2$of each facing half-spaces:

$$f_j = -k(D - u_j).$$

(2)

Here, *k* is the spring constant of the fiber, *D* is the explicit imposed displacement to the rigid half-space with respect to the elastic half-space as seen in Figure 1, and *u* is the local displacement of the surface of the elastic half-space from equilibrium due to the force exercised by all the other fibers in the array. *j* is the index of the fiber. The spatially dependent displacement of the elastic medium owing to this force was calculated by [1929]. He obtains the displacement of the fiber *j* as:

$$u_j = \sum_{i \neq j}^{N} G_{ij} f_i.$$

(3)

The Green's function is:

$$G_i = \frac{1-\nu^2}{\pi E a^2} \int_{-a/2}^{+a/2} \int_{-a/2}^{+a/2} \frac{dx'\,dy'}{|\vec{r}_0 - \vec{r}_i(x', y')|}, \tag{4}$$

$\vec{r}_0$ is the position vector along the interface (i.e., the fault) where the displacement u_j is computed, and $\vec{r}_i$ is the distance vector centered around the contributing fiber *i*. The *x*' and *y*'vectors have their origin at the *i*th fiber. If we measure *r* in units of the discretization *a*, we can find a single parameter controlling the impact of the Green's function. We call this, the reduced Young's modulus [Stormo et al. (2013])]

$$e \equiv \frac{a \cdot E}{L}. \tag{5}$$

In this way, we can re-scale the results of our calculations. The elastic response from the Green's function is solely depending on the value of e, and thus changing the Young's modulus, is equivalent to changing the discretization size a. Note that the system size is =*aL*. Accordingly, the reduced Young's modulus can be written including the system size:

$$e \equiv a^2 E / \mathcal{L} \tag{6}$$

The model is quasi-static. Each fiber has a strength f_j^t at which it breaks irreversibly if the applied force exceeds. The thresholds are following a random distribution. For the sake of this paper, the distribution is uniform. We impose the displacement *D*, calculate the forces on the fibers and find which fiber has the largest f/f^t ratio. We then re-scale *D* so that this fiber is the only fiber with a force higher or equal to its threshold and break. We then redistribute the forces according to Equation 4 following an event-driven algorithm. This is done iteratively until all fibers of the system are broken. From this iterative process, it is possible to calculate the global stress , as a function of the global displacement *D* of the system. The global stress is calculated for each iteration as the sum of force applied on the fibers divided by the area of the system:

$$\Sigma = \frac{1}{a^2 N} \sum_{i=1}^{N} f_i \tag{7}$$

We count time p as the ratio between broken fibers, n, and the total amount of fibers in the system$N$: $p=n/N$. A typical plot of as a function of the damage is given in Figure 2. When reaches its maximum, the system cannot support any additional load. We refer to the damage ratio at this point as the critical damage, p_c.

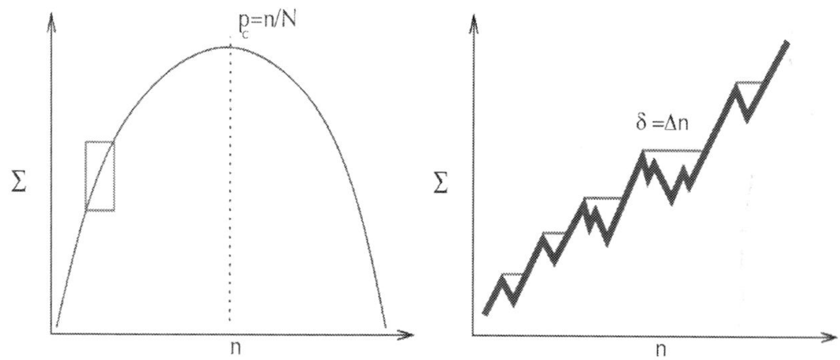

Figure 2: Typical load-displacement curve. Left: Typical *Σ*(*n*) force evolution with 'time' (i.e., number of broken fibers) in the elastic fiber bundle. Right: Zoom of force evolution indicated by the red area. Burst sizes are defined as the number of fibers broken before an increase of global load is obtained. They are illustrated by horizontal red lines.

When a fiber breaks, a new equilibrium corresponding to a new u displacement field is computed, leading to a new load on the surviving fibers. Depending on the reduced Young's modulus e, the redistribution of the load can be diffused or highly localized. The model contains two regimes of different behavior in this regard [Stormo et al. (2013])]. When the Young's modulus is so high that the Green's function is essentially turned off, the model is equivalent to the equal load sharing fiber-bundle model [Daniels (1945])]. We will call this regime the stiff regime. The system undergoes critical damage at p_c^{stiff}. As the system gets softer, a cross-over regime appears as the spatial influence described by the Green's function begins to become relevant compared to the imposed displacement. At the end of this cross-over, we have the soft regime, where the Green's function is completely turned on. This system undergoes critical damage at $p_c^{soft} < p_c^{stiff}$ [Stormo et al. (2013])].

There is no clear value of e separating the stable regimes from the cross-over, but we will use the values for stiff and soft systems that were used by [Stormo et al. (2013]), as they are safely outside the boundaries of the cross-over: $e_{stiff}=32$ and $e_{soft}=7.63\cdot10^{-6}$. The force felt by a fiber is proportional to $u-D$. For stiff systems, $D \gg u$ and the influence of u is negligible. For soft systems$u \sim D$ and the load distribution is highly dependant on the Green's function. When we compare $D$and u for a system with no broken fibers, we get $u=0.0162D$ for e=32 and $u=0.9999D$ for$e=7.63\cdot10^{-6}$.

The Ductile to Brittle Transition

There are several ways to define the difference between ductile and brittle damage. Just as [Amitrano (2003]), we will use the amount of inelastic, i.e., the non-linear deformation the system sustains before critical damage occurs as a measure of ductility. If there is instead a strongly linear load-displacement all the way until failure, we name it a brittle system. We then study the load-displacement curves from the stiff (e=32) and soft ($e=7.63\cdot10^{-6}$) systems shown in Figure 3. On top, we can see the load-displacement for stiff systems. It is clear that the load-displacement curve is highly non-linear until critical damage (indicated by the star). This corresponds to our definition of ductility. On the bottom, we see the same curve for a soft system. In contrast to the stiff system, there seems to be no sign of non-linear deformation before failure (indicated by the star). Thus, we have a strong indication of a ductile to brittle transition as we move from the stiff regime to the soft one.

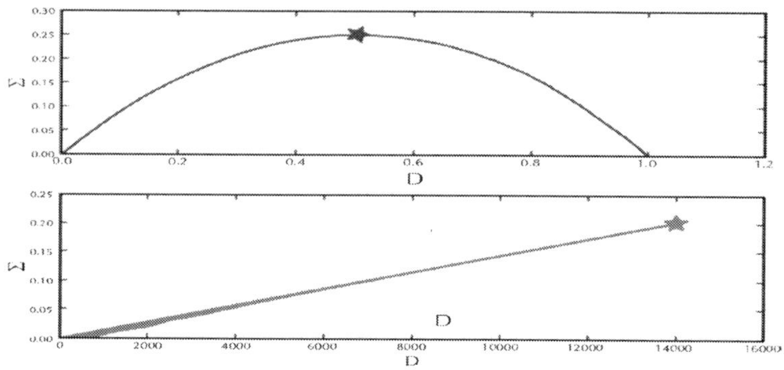

Figure 3: The load as a function of displacement*D*. Top: Load displacement for stiff, ductile systems (e=32). Bottom: Load displacement for soft, brittle systems (e=7.63·10⁻⁶). The point of critical damage is indicated by a star.

In order to see how the brittle-ductile transition affects the microscopic damage process in the two regimes, Figure 4 shows a map of damage at p_c^{soft} and p_c^{stiff} and the distribution of $\langle r^2\rangle^{1/2}$. The red and the blue vertical lines in the $\langle r^2\rangle^{1/2}$ map indicates p_c^{soft} and p_c^{stiff}, respectively. The yellow line on top of the $\langle r^2\rangle^{1/2}$ map shows the ensemble average $|\langle r^2\rangle^{1/2}|$. Figure 5 shows the same for a soft system.

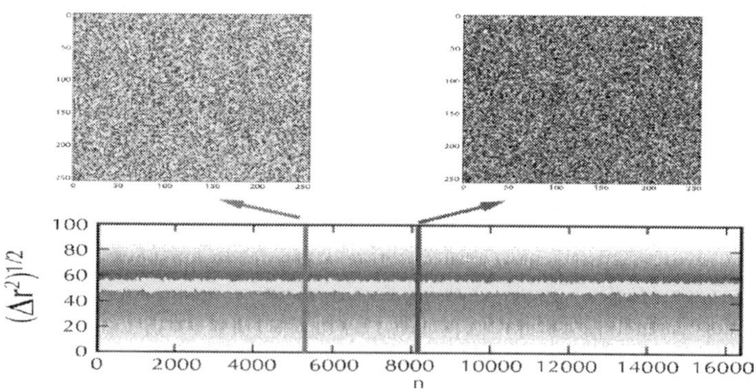

Figure 4: The damage process of stiff (e=32) systems. Top: Damage maps at p_c^{soft} and p_c^{stiff}. The yellow areas indicate intact fibers, while the black ones indicate damaged fibers. Bottom: Distribution

of ?$\Delta r^2\rangle^{1/2}$as a function of fiber-damage n for L=128 (reprinted from [Stormo et al. (2013]) ©2012 American Physical Society). Darker colors indicate higher density. The yellow line shows $|\langle r^2\rangle^{1/2}|$.

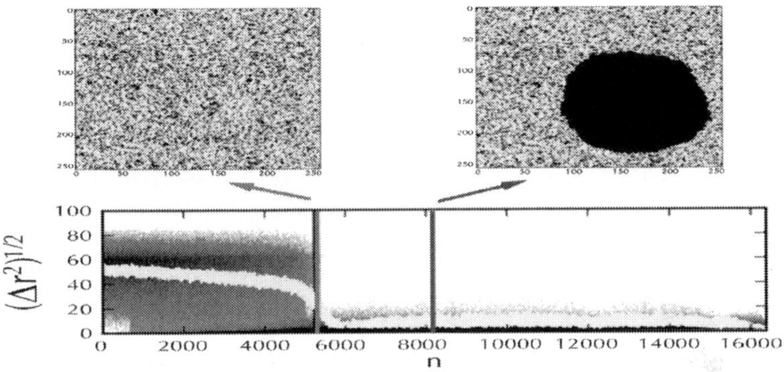

Figure 5: The damage process of soft (e=7.63·10−6) systems. Top: Damage maps at p_c^{soft} and p_c^{stiff}. The yellow areas indicate intact fibers, while the black ones indicate damaged fibers. Bottom: Distribution of $(\Delta r^2)^{1/2}$ as a function of fiber-damage n for L=128 (reprinted from [Stormo et al. (2013]) ©2012 American Physical Society). Darker colors indicate higher density. The yellow line shows $|(\Delta r^2)^{1/2}|$.

In the stiff systems (e=32), the localization of damage is independent from iteration to iteration. The average square distance between two consecutive broken fibers is $\left|\left\langle \Delta r^2 \right\rangle^{1/2}\right| = L/\sqrt{6}$ [Stormo et al. (2013])]. We can see from Figure 4 that the average is slightly above 50, consistent with $L/\sqrt{128} = 52.26$. In soft systems (e_{soft}=7.63·10^{-6}), the influence from the Green's function ensures that the fibers closer to already broken fibers have a larger chance for failure, and thus $|\langle \Delta r^2\rangle^{1/2}|$ dramatically decreases as seen in Figure 4.

The diffuse distribution of damage in the stiff systems and the highly localized damage of the soft systems led to describe the two regimes as ductile and brittle, respectively, see [Anderson (2005]).

RESULTS AND DISCUSSION

Burst from Constant Force Steps

Damage Burst Definition

We now introduce a first definition of burst that is typically employed in quasi-static models in order to compare quasi-static time p to laboratory time. It is assumed that during a simulation, if a fiber breaks and the resulting load redistribution induces more fibers exceeding their force threshold f^t, then the latter fibers are also damaged at the same time. Following the (n)-curve in Figure 2, we can see local maxima where the subsequent fibers fail at a lower load. In laboratory time, we assume the number of fibers:

$$\delta = \Delta n, \tag{8}$$

that break before is increased, fail in an avalanche or burst.

Damage Burst Distribution, b_b

In order to compare our model to the one presented by [Amitrano (2003]), we want to find the distribution of temporal damage bursts. We can then extract a burst distribution exponent b_b-value from our system using the event-size distribution. We recall from Equation 8 that we define the size of an event δ as the number of fibers that are broken for each increase of the load Σ. In Figure 2, we see a zoomed version of $\Sigma(p)$, and we can identify δ. Burst data ends at p_c, as it is the point of maximum Σ. Following [Amitrano (2003]), we assume that the size of events are comparable to the magnitude of acoustic emissions, and we thus get the δ- b_b connection:

$$\log_{10}[N(\delta)] \propto -b \log_{10}[\delta], \tag{9}$$

Where N ( ) is the number of bursts of size or higher.

We run the elastic fiber-bundle model for both the stiff (e=32) regime with diffuse ductile failure, and the soft ($e=7.63\cdot10^{-6}$) regime with localized brittle failure. This is done for both regimes at L= [32,64,128,256].

Figure 6 shows the burst distribution for stiff systems.

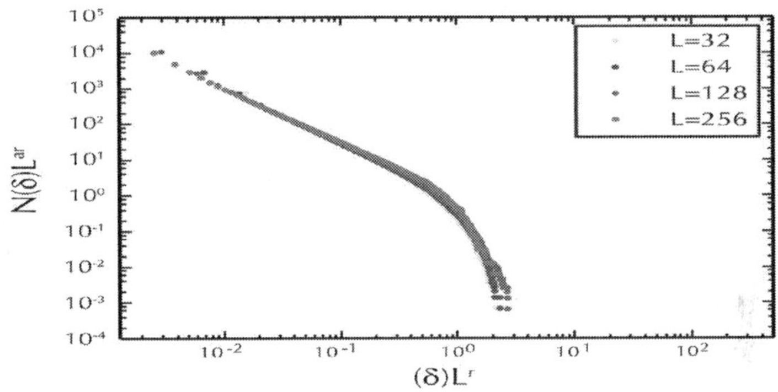

Figure 6: Burst distributions for stiff systems. The cumulative distribution of burst sizes for stiff (e=32) systems. The curves for different system sizes have been collapsed into a single curve. The best collapse has been obtained for α=1.6 and γ=1.2.

The slope of 1.475 for the linear trend for $\delta/L^{\gamma}\leq1$ is consistent with the theoretical 1.5 burst exponent for the equal load sharing fiber bundle model (see [Pradhan et al. (2010]), [Hemmer and Hansen (1992]) and [Pradhan et al. (2005])). This gives us a size-independent stiff b_b-value of 1.475 for L= [32,64,128,256]. We also observe a good collapse of the data for the different size systems using the collapsing method of [Girard et al. (2010]) (Figure 6) and obtained similar values of γ=1.2 and α=1.6.

In contrast from the stiff systems, the soft systems have size-dependent b_b-values (see Figure 7). The slope of the burst distribution gives $b_b(L)$= [2.02,2.23,2.60,3.00] for L=[32,64,128,256], with no clear sign of convergence. As changing L also influences the elastic behavior of the system from Equation 5, we also tested the sole influence of e on the slope of the burst distribution, keeping L constant and taking L=64 (see Figure 8). Results obtained in these simulations are less precise that those obtained for e=32 and $e=7.63\cdot10^{-6}$, but they highlight that the slope is controlled by the value of e. At high e the slope reaches the value 1.475 independently of e as observed previously while the b_b-value goes to higher values as e decreases. It confirms that$e \approx 10^{-2}$ is at the transition between both regimes.

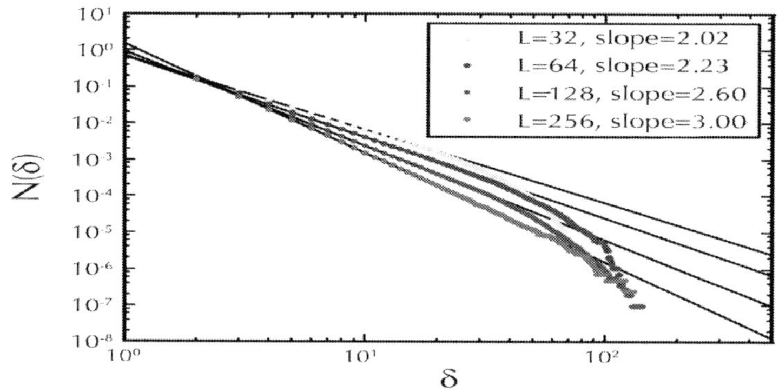

Figure 7: Burst distributions for soft systems. The cumulative distribution of burst sizes for soft (e=7.63·10^{-6}) systems. The black line is a guide to the eye for the given slope of the distribution. The b_b-values extracted are b_b (L)= [2.02,2.23,2.60,3.00].

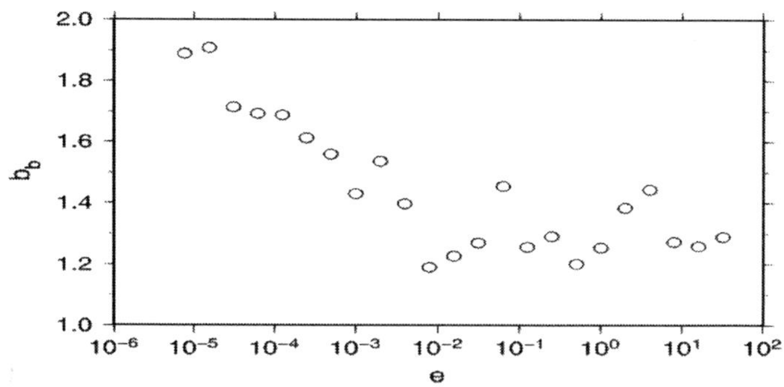

Figure 8: b_b-values of damage burst distribution.b_b-values obtained from the slope of the cumulative distribution of damage burst sizes for various e moduli. The system size is set to L=64. Each point is an average over 400 simulations.

Burst from Thresholding the Local Fracture Front Velocity

Velocity Burst Definition

Another approach to characterize the burst activity of the model for a comparison with earthquake seismicity is to follow the analysis proposed by [Grob et al. (2009]). Here, we extend this approach from a single front monitoring, initially proposed in [Måløy et al. (2006]) to a full interface with multiple active fronts at the same time. In order to first calculate the velocity of the fracture front, we use the concept of waiting time matrix, *W*, introduced by [Måløy et al. (2006]). The waiting time matrix, *W*, is a matrix with dimension $L \times L$, storing the number of times a fiber is part of a fracture front. Initially, every element of *W* is set to 0. At every time step, the positions of the fracture fronts are identified, and the corresponding elements of *W* are increased by one.

We define any surviving fiber that neighbors a damaged one as a part of an evolving fracture front.

When the fracture is complete, we create the velocity matrix by taking the inverse of the elements of *W* times the discretization length:

$$V(x,y) = \frac{a}{W(x,y)}. \tag{10}$$

where $V(x,y)$ is the velocity at which the fracture front passes trough element (x,y).

Finally, in order to identify velocity bursts, we introduce the threshold level *C* as in [Måløy et al. (2006]). We first calculate the average velocity of the fracture front, $\langle V \rangle$. We then go through the velocity matrix and create the velocity cluster map by setting every element of $V(x,y)$ above $C \cdot \langle V \rangle$ to 1 and the rest to 0. We then go trough this cluster map and identify the cluster size distribution using a Hoshen-Kopelman algorithm [Hoshen et al. ([1978])]. By doing this, we can identify the cluster size distribution $p(s,C)$, where *s* is the size of any given cluster. Example of velocity map and velocity cluster map are given in Figure 9 for stiff systems and in Figure 10 for soft systems. We see that clustering emerges for soft systems with the highest velocities.

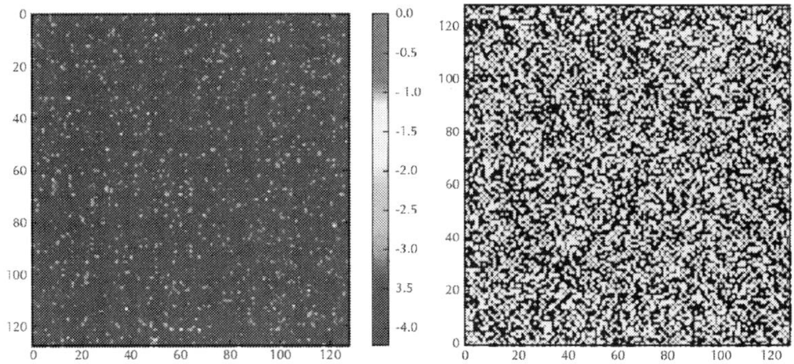

Figure 9: Velocity map for stiff systems. L=128, left: Map of log10[$V(x,y)$] for a stiff system. The average velocity is $\langle V\rangle$=0.000147. Right: The corresponding C=2 cluster-map. We identify the velocity cluster by setting all elements above $C\cdot\langle V\rangle$ to one (yellow) and below to zero (black).

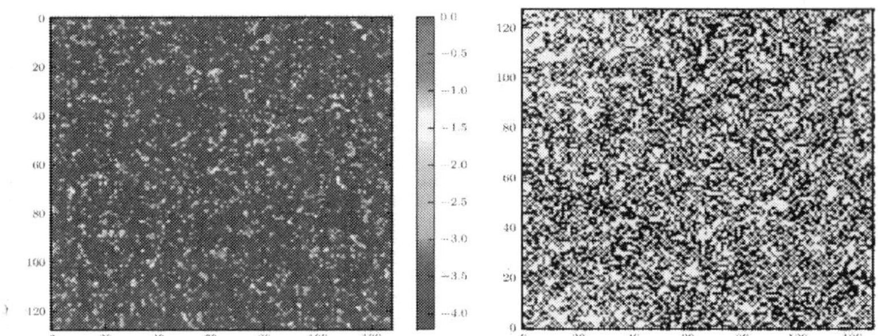

Figure 10: Velocity map soft systems. L=128, left: Map of log10[$V(x,y)$] for a soft system. The average velocity is $\langle V\rangle$=0.000187. Right: The corresponding C=2 cluster-map. We identify the velocity cluster by setting all elements above $C\cdot\langle V\rangle$ to one (yellow) and below to zero (black).

Velocity Burst Distribution, b_v

We calculate the cluster distribution $p(s,C)$ for both regimes. This distribution is given in Figure 11for stiff systems and in Figure 12 for soft systems.

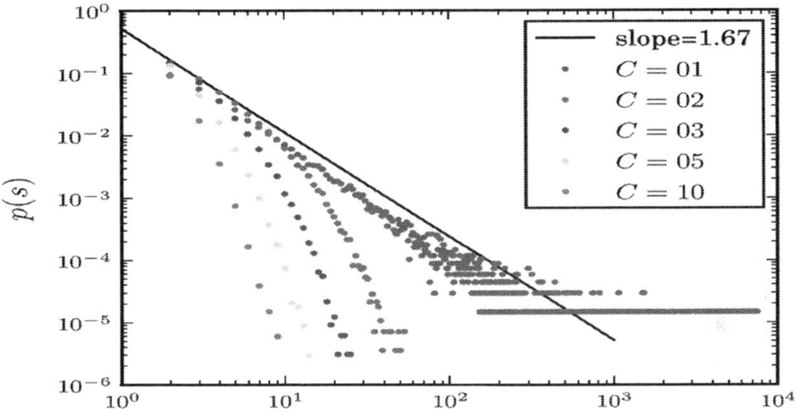

Figure 11: Distribution of velocity clusters for stiff systems. The velocity cluster distribution $p(s,C)$ for L=128, C= [1,2,3,5,10]. The black 5/3 slope is added for comparison.

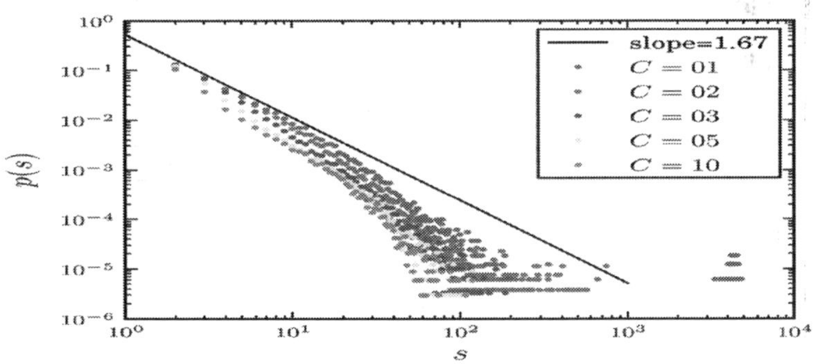

Figure 12: Distribution of velocity clusters for soft systems. The velocity cluster distribution $p(s,C)$ for L=128, C= [1,2,3,5,10]. The black 5/3 slope is added for comparison.

In order to compare our model to seismic activity, we will follow the analysis of Grob et al. ([2009]). A link is made between the area of the velocity clusters and the seismic moment M_0 following [Kanamori and Anderson (1975]):

$$M_0 = \mu s d. \tag{11}$$

Where μ is the shear modulus and d is the slip offset. Since our model does not undergo slip but only mode I opening, we assume d to be constant. Accordingly, the probability density for seismic moment follows the power law:

$$p(M_0) \propto M_0^{-1-B}, \tag{12}$$

where B has been found to be stable around 2/3 for a large number of seismic regions [Scholz CH (2002])].

In both Figures 11 and 12, we show a 5/3=2/3+1 slope for comparison. We can see that for small$C$-values (pink color), the cluster distribution is consistent with B=2/3 for both regimes in particular for small events. From [Scholz CH (2002]), we get the connection to the Gutenberg-Richter distribution, as b=3/2B. We thus obtain for the front advance distribution, b_v=1 for both ductile and brittle systems. However, a clear finite size effect is visible for soft systems and large events. As Cbecomes larger, the difference between both stiff and soft regimes increases. In particular, we obtain an artificially large slope (large b-values) for stiff systems (Figure 11) for large C-values. This is not the case for soft systems for which the C-value only changes the pre-factor (i.e., a-value) but not the general behavior of the distribution. This observation suggests that the change in b-value can be related to the *a posteriori* clipping level and not to a physical parameter of the model. In this case, b-value changes are more related to recording instrumentation rather than different physical processes.

When we look at the velocity clusters, we can see that there is a possible $b_v \approx 1$ value for both ductile and brittle systems. However, for the system sizes currently available, the large clusters get rare as we increase C. This happens faster for the stiff systems, as the velocity distribution is narrower (Figure 9) than for the soft systems (Figure 10). With this result, we extend to a new configuration (multiple fronts), the results obtained for a single directional front propagation reported by both [Måløy et al. (2006]), [Grob et al. (2009]), and [Tallakstad et al. (2011]).

The *C*-level is the minimum local crack velocity that we have in a velocity cluster. As *C* increase, the average local velocity of the clusters is larger. If we compare to shear rupture, an increase of *C* is analogous to a rising of the average slip velocity. In that sense, it will select only events with large velocities or amplitudes if measured with a velocimeter. This assumes a direct link between rupture velocity along the fault and wave magnitudes at the seismometer. Another discrepancy with our model and the seismic activity is the fact that our model is quasi-static and not dynamic (i.e., no dynamical rupture propagation). Thus, it is not straightforward to compare with dynamic rupture and dynamic effects. In order to get a firmer connection to the seismic activity, the model has to be further developed.

Our model displays the same type of change in b_b-value during ductile to brittle transition, i.e., a transition from diffuse to localized damage in the system, as described by [Amitrano (2003]). In the brittle regime, we get b_b-values that are twice the *b*-values observed during the hydraulic injections in the EGS experiments (average of 1.58 during injections, 1.15 after shut-in at Basel,[Bachmann et al. (2011])) (average of 1.29, ranging from 1.0 to 1.7 in the GPK2 site in Soultz-sous-Forêts, [Cuenot et al. (2008])]).

In the analysis of [Cuenot et al. (2008]), they found that when they increased the injection rate from 40 to 50 ls $^{-1}$ they initially saw a drop in the *b*-value. Eventually it recovered, before it climbed to a new peak. They speculate that this seemed to come from the fact that the weak points in the fault were already damaged, and that the strong points had to break for the water to seep trough to the rest of the network. Breaking the strong points would release more energy giving an initial drop in the *b*-value. When the fluids finally were able to seep through to the rest of the network, a lot more weak points were available. Thus, a significant increase in *b* was to be expected.

In our model, we see a similar increase in the b_b-value when we increase the number of fibers in the system (*L*). In the fiber bundle model, the weakest bonds have the highest risk of failure. As the system becomes larger, the average distance between the weak points increases due to the uncorrelated nature of the threshold distribution. When we compare small systems with fewer fibers, the build-up of stress concentration will be much denser. Thus, the chance of a failure triggering an avalanche nearby is higher. This explains why our b_b-value increases with system size. This could be linked to the Cuenot's argument for the *b*-value fluctuation.

The model does come with its share of limitations though. First of all, we do not get the exact same b_b-values as seen in the hydraulic injections. And even though we can see a change of b_b-value, the transition described by [Cuenot et al. (2008]) cannot at this stage be addressed if the change comes from the opening of the fracture network, as our model is dealing with a single fault. Connecting different samples and activating them as the fluid spreads through the system might be a way of addressing this issue.

Secondly, as illustrated in Figure 1, the load imposed on our fiber bundle is external. This is in contrast to the internal pressure experienced by the faults during the hydraulic injections. Rules for applying the development of hydro-pressure in the weak plane between the elastic half-spaces should be developed to account for this. This should not change the burst distribution in the stiff regime as it is controlled by the disorder, but might have an effect on the soft systems.

A third limitation of our model in connection to the EGS experiments, is the fact that the model is describing a pure mode I loading. This is in contrast to the typical slip faulting in earthquakes. Fortunately, the energy released for mode I, II, and III fracture only differ by a prefactor as described by [Schmittbuhl et al. (2003]). Thus, the transfer of load and the shape of the damage burst distribution should be applicable to those cases as well.

In our model, the length *a* (see Equation 5) refers to the distance between two neighboring fibers. In this sense, the quantity, $1/a$ can be understood as the number of fiber per unit length (or as a fiber linear density). Then, for a fixed system size, $/a=L$ represents the space density of fibers which can be considered as the density of contact points between the facing surfaces of the fracture. In a geological context these contact points are referred as asperities along the fault plane. Changing the asperity density *L* along the fault plane (i.e., the heterogeneity of the interface), implies changing the important parameter of the model: $e=(a \cdot E)/L$ (from Equation 5). When e is small, the system is considered as soft, localization of the plastic deformation (i.e., fracture) is observed and *b*-values are untypical. When e is large, the system is stiff, deformation is diffused and *b*-values are more typical. It is of interest to see the Young *E* might have a inverse effect on e compared to the asperity density *L*. Elastically stiff systems (i.e., large *E*) are similar to a small density of asperities *L* in particular if the asperity size gets close to the system size: $a\rightarrow$. Stiff fiber-bundle models can then be viewed as systems with large scale heterogeneities along the interface (i.e., large *a* and small *L*) for which the damage zone is notably extended.

The main difficulty to relate our model to natural fault system is to assess the degree of heterogeneity of the interface at depth event in the context of geothermal reservoir. If we were able to resolve such properly, we could then make a direct comparison between the *b*-values encountered in various interfaces with the results of our model. Although we have limited knowledge of the fault heterogeneity, we propose two observations that is linked to this feature and would help to draw some link between our model and natural reservoir. First, it has been shown that the contact area along a rough fault, is typically related linearly, at first order, to the normal stress along the fault [Batrouni et al. ([2002a]), Hyun et al. ([2004])]. Accordingly, the asperity size *a* will tend to the system size as normal stress is increased. The system is then expected to become effectively stiffer (i.e., increase of e) as the normal stress increases along the fault. If so the *b*-value is expected to decrease the normal stress increases (see Figure 8). Second, one could use the out of plane distribution of seismicity as a function of the fault thickness. As shown for faults in California, this distribution can be related to the stress variations owing to the fault roughness and thus can be interpreted as a quantification of the heterogeneity of the interface [Powers and Jordan (2010])]. In this sense, it would then be directly related to the *L*parameter introduced in our system. Provided that earthquake locations are of sufficient accuracy to estimate precisely such a distribution, it would then make a link between the parameters of our model and observable quantities.

CONCLUSIONS

We have found that the elastic fiber bundle model shows promising features to understand the mechanical origin of the *b*-value fluctuations during stimulation of deep geothermal reservoirs. We proposed two types of events in our model that can be compared to seismic events: a classical measure, the damage burst, i.e., the number of broken fibers during an elementary rise of the loading stress, and a new measure, the velocity burst, i.e., the size of a front velocity cluster during its propagation. We show that both event definitions lead to different event distributions. The damage event distribution has a *bb* value that is influenced by the normalized *e* modulus: changing at small *e* (soft systems) but constant at large *e* (stiff systems). The second definition based on the front velocity produces distributions with a b_v value that is evolving differently. We have shown the feasibility of connecting the velocity cluster distribution to the distribution of seismic moment M_0 of earthquakes. The main change of the b_v-value is coming from the clipping level and not from the transition from stiff to soft (or ductile to brittle). It is then related to the observation sensitivity. It suggests that *b*-value changes could be related to measurement evolution.

AUTHORS' CONTRIBUTIONS

AS carried out the numerical study. OL participated in the design of the study and in the statistical analysis. JS conceived of the study and participated in its design and coordination. All authors read and approved the final manuscript.

ACKNOWLEDGEMENTS

This work has been published under the framework of the LABEX G-EAU-THERMIE-PROFONDE ANR-11-LABX-0050 and benefits from a funding from the state managed by the French National Research Agency as part of the 'Investments for the Future' program. We would like to thank Alex Hansen and Knut Skogstrand Gjerden for ideas and insights to the elastic fiber-bundle model. We also thank NOTUR for the allocation of computer time. Two anonymous reviewers are warmly acknowledged for their constructive comments.

REFERENCES

1. Amitrano D: Brittle-ductile transition and associated seismicity: experimental and numerical studies and relationship with the *b* value. *J Geophys Res Earth 1978–2012* 2003, 108(B1):2044.
2. Amitrano D, Grasso J-R, Hantz D: From diffuse to localised damage through elastic interaction. *Geophys Res Lett* 1999, 26(14):2109-2112.
3. Anderson, TL (2005) Fracture mechanics: fundamentals and applications. CRC press.
4. Aochi H, Poisson B, Toussaint R, Rachez X, Schmittbuhl J: Self-induced seismicity due to fluid circulation along faults. *Geophys J Int* 2014, 196(3):1544-1563.
5. Bachmann CE, Wiemer S, Woessner J, Hainzl S: Statistical analysis of the induced Basel 2006 earthquake sequence introducing a probability-based monitoring approach for enhanced geothermal systems. *Geophys J Int* 2011, 186(2):793-807.
6. Baisch S, Vörös R, Rothert E, Stang H, Jung R, Schellschmidt R: A numerical model for fluid injection induced seismicity at Soultz-sous-Forêts. *Int J Rock Mech Mining Sci* 2010, 47(3):405-413.
7. Batrouni, GG, Hansen A, Schmittbuhl J (2002a) Elastic response of rough surfaces in partial contact. Europhys Lett 60(5): 724–730.
8. Batrouni, GG, Hansen A, Schmittbuhl J (2002b) Heterogeneous interfacial failure between two elastic blocks. Phys Rev E 65(3): 036126.
9. Cuenot N, Dorbath C, Dorbath L: Analysis of the microseismicity induced by fluid injections at the EGS site of Soultz-sous-Forêts (Alsace, France): implications for the characterization of the geothermal reservoir properties. *Pure Appl Geophys* 2008, 165(5):797-828.
10. Daniels HE: The statistical theory of the strength of bundles of threads. I. *Proc R Soc London. Series A Math Phys Sci* 1945, 183(995):405-435.

11. Dorbath L, Cuenot N, Genter A, Frogneux M: Seismic response of the fractured and faulted granite of Soultz-sous-Forêts (France) to 5 km deep massive water injections. *Geophys J Int* 2009, 177(2):653-675.
12. Enescu B, Ito K: Some premonitory phenomena of the 1995 Hyogo-Ken Nanbu (Kobe) earthquake: seismicity, *b*-value and fractal dimension. *Tectonophysics* 2001, 338(3):297-314.
13. Girard L, Amitrano D, Weiss J: Failure as a critical phenomenon in a progressive damage model. *J Stat Mech Theory Exp* 2010, 01:P01013.
14. Gischig VS, Wiemer S: A stochastic model for induced seismicity based on non-linear pressure diffusion and irreversible permeability enhancement. *Geophys J Int* 2013, 194(2):1229-1249.
15. Grob M, Schmittbuhl J, Toussaint R, Rivera L, Santucci S, Måløy KJ: Quake catalogs from an optical monitoring of an interfacial crack propagation. *PAGEOPH* 2009, 166(5–7):777-799.
16. Gutenberg B, Richter CF: Frequency of earthquakes in California. *Bull Seismological Soc Am* 1944, 34(4):185-188.
17. Hakimhashemi AH, Yoon JS, Heidbach O, Zang A, Grünthal G: Forward induced seismic hazard assessment: application to a synthetic seismicity catalogue from hydraulic stimulation modelling. *J Seismol* 2014, 18(3):671-680.
18. Hansen A, Schmittbuhl J, Batrouni GG, De Oliveira FA: Normal stress distribution of rough surfaces in contact. *Geophys Res Lett* 2000, 27:3639-3643.
19. Hemmer PC, Hansen A: The distribution of simultaneous fiber failures in fiber bundles. *J Appl Mech* 1992, 59(4):909-914.
20. Holliday JR, Turcotte DL, Rundle JB: A review of earthquake statistics: fault and seismicity-based models, ETAS and BASS. *Pure Appl Geophys* 2008, 165(6):1003-1024.
21. Hoshen J, Kopelman R, Monberg EM: Percolation and cluster distribution. II. layers, variable-range interactions, and exciton cluster model. *J Stat Phys* 1978, 19(3):219-242.
22. Hyun S, Pei L, Molinari J-F, Robbins MO: Finite-element analysis of contact between elastic self-affine surfaces. *Phys Rev E* 2004, 70(2):026117.

23. Jaeger JC, Cook NGW, Zimmerman R: *Fundamentals of rock mechanics*. Backwell Publishing, Australia; 2009.
24. Kanamori H, Anderson DL: Theoretical basis of some empirical relations in seismology. *Bull Seismol Soc Am* 1975, 65(5):1073-1095.
25. Kwiatek G, Plenkers K, Nakatani M, Yabe Y, Dresen G: JAGUARS Group: Frequency-magnitude characteristics down to magnitude-4.4 for induced seismicity recorded at Mponeng Gold Mine, South Africa. *Bull Seismological Soc Am* 2010, 100(3):1165-1173.
26. Love AEH: The stress produced in a semi-infinite solid by pressure on part of the boundary. *Phil Trans R Soc A* 1929, 228:377.
27. Måløy K, Santucci S, Schmittbuhl J, Toussaint R: Local waiting time fluctuations along a randomly pinned crack front. *Phys Rev Lett* 2006, 96(4):045501.
28. Powers PM, Jordan TH: Distribution of seismicity across strike-slip faults in California,. *J. Geophys. Res.* 2010, 115:B05305. doi:10.1029/2008JB006234
29. Pradhan S, Hansen A, Hemmer PC: Crossover behavior in burst avalanches: signature of imminent failure. *Phys Rev Lett* 2005, 95(12):125501.
30. Pradhan S, Hansen A, Chakrabarti BK: Failure processes in elastic fiber bundles. *Rev Modern Phys* 2010, 82(1):499.
31. Rundle JB, Turcotte DL, Shcherbakov R, Klein W, Sammis C: Statistical physics approach to understanding the multiscale dynamics of earthquake fault systems. *Rev Geophys* 2003, 41(4):1019.
32. Sausse J, Dezayes C, Dorbath L, Genter A, Place J: 3D model of fracture zones at Soultz-sous-Forêts based on geological data, image logs, induced microseismicity and vertical seismic profiles. *Comptes Rendus Geosci* 2010, 342(7):531-545.
33. Schmittbuhl J, Delaplace A, Perfettini H, Vilotte JP: Måløy KJ: Slow crack propagation and slip correlations. *PAGEOPH* 2003, 160(5–6):961-976.
34. Scholz CH: The frequency-magnitude relation of microfracturing in rock and its relation to earthquakes. *Bull Seismological Soc Am* 1968, 58(1):399-415.

35. Scholz CH: *The mechanics of earthquakes and faulting*. Cambridge University Press, Cambridge, UK; 2002.
36. Schorlemmer D, Wiemer S, Wyss M: Variations in earthquake-size distribution across different stress regimes. *Nature* 2005, 437(7058):539-542.
37. Stormo A, Gjerden KS, Hansen A: Onset of localization in heterogeneous interfacial failure. *Phys Rev E* 2013, 86:025101.
38. Tallakstad KT, Toussaint R, Santucci S, Schmittbuhl J, Måløy KJ: Local dynamics of a randomly pinned crack front during creep and forced propagation: an experimental study. *Phys Rev E* 2011, 83:046108.
39. Turcotte DL, Newman WI, Shcherbakov R: Micro and macroscopic models of rock fracture. *Geophys J Int* 2003, 152(3):718-728.
40. Wyss M: Towards a physical understanding of the earthquake frequency distribution. *Geophys J R Astronomical Soc* 1973, 31(4):341-359.
41. Yoon, JS, Zang A, Stephansson O (2014) Numerical investigation on optimized stimulation of intact and naturally fractured deep geothermal reservoirs using hydro-mechanical coupled discrete particles joints model. Geothermics.
42. Zang A, Oye V, Jousset P, Deichmann N, Gritto R, McGarr A, Majer E, Bruhn D: Analysis of induced seismicity in geothermal reservoirs - an overview. *Geothermics* 2014, 52:6-21.

Chapter 5

Retrospective Analysis of Uncertain Eruption Precursors at La Soufrière Volcano, Guadeloupe, 1975–77: Volcanic Hazard Assessment using a Bayesian Belief Network Approach

Thea K Hincks[1], Jean-Christophe Komorowski[2], Stephen R Sparks[1], and Willy P Aspinall[1, 3]

[1]School of Earth Sciences, University of Bristol, Queen's Road, Bristol BS8 1RJ, UK

[2]Institut de Physique du Globe de Paris, CNRS UMR 7154, PRES Sorbonne Paris-Cité1 rue Jussieu, 75238 Paris, Cedex 05, France

[3]Aspinall & Associates, Cleveland House, High Street, Tisbury SP3 6HF, UK

ABSTRACT

Background

Scientists monitoring active volcanoes are increasingly required to provide decision support to civil authorities during periods of unrest. As the extent and resolution of monitoring improves, the process of jointly interpreting multiple strands of indirect evidence becomes increasingly complex. Similarities with uncertainties in medical diagnosis suggest a formal evidence-based approach, whereby monitoring data are analysed synoptically to provide probabilistic hazard forecasts. A statistical tool to formalize such inferences is the Bayesian Belief Network (BBN). By explicitly representing conditional dependencies between the volcanological model and observations, BBNs use probability theory to treat uncertainties in a rational and auditable manner, as warranted by the strength of the scientific evidence. A retrospective analysis is given for the 1976 Guadeloupe crisis, using a BBN to provide inferential assessment of the state of the evolving magmatic system and probability of incipient eruption. Conditional dependencies are characterized quantitatively by structured expert elicitation.

Results

Analysis of the available monitoring data suggests that at the height of the crisis the probability of magmatic intrusion was high, in accordance with scientific thinking at the time. The corresponding probability of magmatic eruption was elevated in July and August 1976 and signs of precursory activity were justifiably cause for concern. However, collective uncertainty about the future course of the crisis was also substantial. Of all the possible scenarios, the most likely outcome evinced by interpretation of observations on 31 August 1976 was 'no eruption' (mean probability 0.5); the chance of a magmatic eruption/blast had an estimated mean probability of ~0.4. There was therefore no evidential basis for asserting one scenario to be significantly more likely than another.

Conclusions

Our analysis adds objective probabilistic expression to the volcanological narrative at the time of the 1976 crisis, and demonstrates that a formal evidential case could have supported the authorities' concerns about public safety and decision to evacuate. Revisiting the episode highlights many challenges for modern, contemporary decision making under conditions of considerable uncertainty, and suggests the BBN is a suitable framework for marshalling multiple, uncertain observations, model results and interpretations. The formulation presented here can be developed as a tool for ongoing use in the volcano observatory.

BACKGROUND

During a volcanic crisis, decisions typically have to be made with limited information and high uncertainty, on short time scales. The primary goal is to minimise loss and damage from any event, but social and economic loss resulting from false alarms and evacuations must also be considered (Woo 2008). Although it is not the responsibility of the scientist to call an evacuation or manage a crisis, there is an increasing requirement to assess risks and present scientific information and associated uncertainties in ways that enable public officials to make urgent evacuation decisions or other mitigation policy choices.

In the interests of safety for both exposed populations and scientists in the field, Aspinall et al. (2003) stress the need for a robust and defensible evidence-based approach to hazard and risk assessment. Although applied in medicine (Sackett et al. 1996), formal evidence-based decision-making is not common practice in volcanology (or indeed other areas of natural hazard and risk assessment). This paper aims to demonstrate the practicality and utility of this approach, particularly in situations with uncertain scientific information, limitations of understanding, and intrinsically unpredictable outcomes. The process of identifying and attempting to quantify sources of uncertainty (both epistemic and aleatory) can be very informative, from all key perspectives (scientists, risk assessors and stakeholders).

Scientific disagreements and political contretemps surrounded the 1976 volcanic crisis of La Soufrière volcano, Guadeloupe (Fiske,

1984; Feuillard et al. 1983) [a]. These controversies and subsequent costly evacuation highlighted the need for a more structured and transparent approach to the use of scientific advice in volcanic hazard assessment (Aspinall and Woo 1994). A key question that needed to be answered in 1976 was: "Is the current state of unrest magmatic in origin, and if so, what is the probability of an explosive volcanic eruption?" An additional question was: "If an explosive eruption occurs, what is the probability that it occurs at the onset or very near the onset of the eruptive activity?" Indeed, the predominant working hypothesis - which influenced the scientific attitude of many and drove the political management of the crisis - was that if there were to be paroxysmal explosive activity then it was more likely to occur at the onset of an eruption, as in the tragic 1902 eruption of La Montagne Pelée, Martinique, rather than later.

[a]One of us, WPA, was present at various times in Guadeloupe in 1976, participating in monitoring activities. Another, JCK, was Director of the Guadeloupe Volcano Observatory (IPGP) from 1997 to 2001.

These questions will be just as pressing if, in the near future, the signs of volcanic unrest manifested since 1992 in Guadeloupe (Komorowski et al. 2005; Villemant et al. 2005) were to escalate. The public clashes in 1976 left a legacy of loss of trust in scientists and authorities there. Any future crisis will need to be managed with care and a transparent and robust approach to information sharing if trust is to be fully restored. Villemant et al. (2005) remark that it remains very difficult to interpret monitoring data in terms of deep volcanic processes, for example to identify or differentiate between magmatic activity and purely hydrothermal activity. At the time of the 1976 crisis, knowledge of the style and magnitude of past eruptive activity of La Soufrière was limited, and multi-parameter monitoring data (e.g. seismicity, deformation, fluid geochemistry) inconclusive and, even jointly, insufficient to reliably inform hazard mitigation decisions.

Following the approach of Aspinall et al. (2003), we use a Bayesian Belief Network (BBN) to interpret jointly the observational evidence available in 1976, and make inferences about the key volcanic states and processes at the time. The output is a retrospective probabilistic forecast that communicates the perceived level of hazard and associated scientific uncertainty. BBNs have been widely applied in engineering and medical decision support systems (Spiegelhalter et al. 1993). The Bayesian methodology is also becoming increasingly used for decision support in natural hazard assessments, including flood risk, terrain analysis and water quality (see, e.g. Molina et al. 2005; Stassopoulou et al. 1998; Borsuk et al. 2003, respectively); and operational risk (Cowell et al. 2007; Neil et al. 2005).

The graphical nature of the BBN makes it an efficient and intuitive means for describing a complex, multi-faceted system. Causal relationships are easily visualised, and can be presented in a compact and easily communicated format (see Methods). Measures such as mutual information (the strength of the relationship between a pair of nodes – see Methods) and entropy can also be used to assess the relative evidential value of individual observations and identify where further data or research may improve the hazard forecast or help reduce uncertainty. All these features should aid communication between volcanologists, risk assessors and stakeholders - critical to the successful management of any volcanological unrest situation. Complementary approaches include logic or event trees (e.g. Newhall and Hoblitt 2002; Marzocchi et al. 2004; Marzocchi et al. 2008; Sobradelo and Martí 2010) - these are generally designed to capture a sequence of events and observations rather than describe the primary components and process interactions of the system; however, the basic probability calculus is largely the same.

Volcanic Record of La Soufrière, Guadeloupe

La Soufrière is the most recently active part of the composite, andesitic La Grande Découverte-Soufrière volcanic (GDSV) complex, located in southern Basse-Terre, Guadeloupe (Figures 1 and 2). Comprehensive descriptions of the geological setting and volcanic activity (Komorowski et al. 2005; Boudon et al. 2008; Samper et al. 2009; Legendre et al. 2010; Legendre 2012) are summarised briefly here. Past activity has ranged from effusive fissure eruptions, explosive magmatic and phreatic episodes, and major sector collapse events (Boudon et al. 1987; Carlut et al. 2000; Komorowski et al. 2008a; Samper et al. 2007) (Table 1).

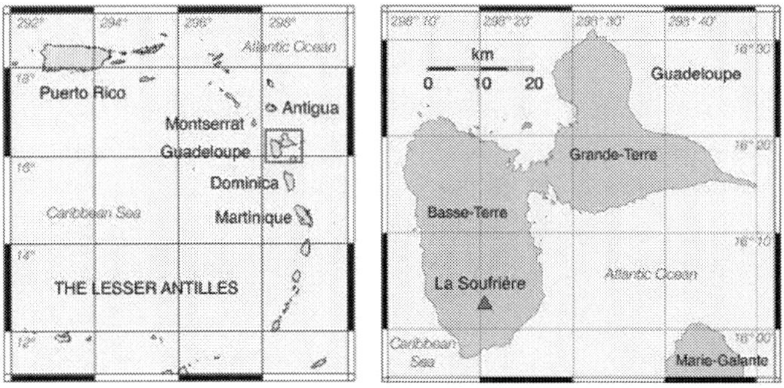

Figure 1: Map of Guadeloupe Location of La Soufrière volcano, Guadeloupe, in the Lesser Antilles island arc, eastern Caribbean.

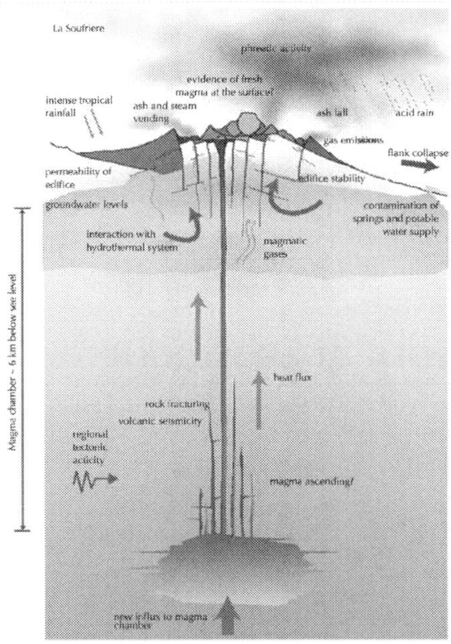

Figure 2: Schematic diagram of La Soufrière volcano. This illustration shows the types of process and phenomena identified at La Soufrière. The BBN in Figure 3 is a development of this simple conceptual volcano model (focussing on the observations available at the time of the crisis and associated volcanic processes) and aims to capture how the key elements of this system are linked.

Table 1: Eruptive chronology of La Soufrière in the last 9000 years, with calibrated radiocarbon dates (CE: Common Era and BCE: before Common Era) from Boudon et al. (1988); Komorowski et al. (2005; 2008); Boudon et al. (2007and 2008); Siebert and Simkin (2002–2011)

Start date	Type	Description
6535 BCE	E	Edifice collapse - not magmatic, not explosive
4000 BCE?	M	VEI 2 explosive Strombolian
3600 BCE?	M	VEI 2 explosive Vulcanian
3360 BCE	M	VEI 3 magmatic dome eruption, possibly explosive
2400 BCE?	E	Edifice collapse - not magmatic, not explosive
1625 BCE	M	VEI 3–4 explosive magmatic with edifice collapse and blast (possible cryptodome?)
1400 BCE	M	VEI 3–4 explosive magmatic with edifice collapse and blast (cryptodome)
1065 BCE?	E	Edifice collapse - not magmatic, not explosive
980 BCE	M	VEI 3 magmatic dome eruption, possibly explosive
465 BCE	M	VEI 3 explosive magmatic dome eruption with edifice collapse and blast
310 CE	M	VEI 2 explosive Strombolian
605 CE	E	Edifice collapse - not magmatic, not explosive
1530 CE	M	VEI 2–3 explosive Subplinian and dome magmatic eruption with edifice collapse
1635 CE ?	M	VEI 2 explosive magmatic, possibly Vulcanian
1690 CE	P	VEI 1 Phreatic - not magmatic but explosive (Komorowski et al.2005)
1797 CE	P	VEI 1 Phreatic - not magmatic but explosive (Komorowski et al.2005)
1812 CE	P	VEI 1 Phreatic - not magmatic but explosive (Komorowski et al.2005)
1836 CE	P	VEI 1 Phreatic - not magmatic but explosive (Komorowski et al.2005)
1956 CE	P	VEI 1 Phreatic - not magmatic but explosive (Komorowski et al.2005)

1976 CE	F	VEI 1 failed (still-born) magmatic explosive (Komorowski et al.2005)

Events are classified as non-magmatic, non-explosive edifice collapses (E); magmatic explosive (M); non magmatic but explosive phreatic events (P), or failed magmatic (F), as in 1976. A question mark indicates the eruption date is uncertain. The last confirmed major magmatic eruption of La Soufrière was 1530 AD (Boudon et al. 2008and Komorowski et al. 2008a 2008b). Phreatic eruptions (identified from historic reports as well as stratigraphic evidence) appear to be more frequent in the last 400 years; however there is significant bias in the catalogue. Evidence of less energetic eruptions before settlement on the island is likely to have been missed, destroyed during larger magmatic events, or poorly preserved in the stratigraphic record. This is also the case for low magnitude magmatic eruptions (VEI 2) as shown by Legendre (2012).

Hincks et *al*

Hincks et *al Journal of Applied Volcanology* 2014 3:3, doi: 10.1186/2191-5040-3-3

The GDSV complex consists of three major composite andesitic volcanoes formed over the last 445 ky: the Grande-Découverte volcano, the Carmichaël volcano, and the most recent La Soufrière volcano (Boudon et al. 1988; Komorowski et al. 2005; Samper et al 2009). Although the Grande-Découverte phase (445 to 42 ka) was dominated by effusive activity it was interrupted by at least three major caldera-forming explosive eruptions approximately at 140, 108, and 42, ka . The 42, ka caldera eruption marks the end of the Grande-Découverte phase (Boudon et al. 1988; Komorowski et al. 2005). This was followed by effusive and explosive activity, forming a new edifice in the caldera. The Carmichaël phase (35 to 11.5 ka) was dominated by effusive to explosive dome-growth and ended with a sector collapse and associated laterally-directed explosions without a magmatic component. The Soufrière phase (last 9 ky) has been characterized by recurrent effusive to explosive dome-growth and up to eight sector collapse events (Komorowski et al. 2005; Boudon et al 2007; Komorowski et al. 2008b) with major events about 8 and 3 ka (Boudon et al. 2007). Several of the sector-collapses were associated with lateral blasts (2–5 events) and a magmatic component (Boudon et al. 1987; Komorowski et al. 2008b).

At least three low to moderate (VEI 2–3) explosive eruptions occurred in the last 9000 years, and three or more much larger VEI 4–5 explosive eruptions in the period from 11.5 to 42.5 ka. The last significant explosive magmatic eruption (VEI 2–3, 1530 AD) produced a multi-stage sequence: a debris avalanche from sector collapse; pumice and scoria fallout from a short-lived subplinian convective column; column collapse scoria pyroclastic density currents, and an andesite lava dome that ended the eruption (Boudon et al. 2008; Komorowski et al. 2008a). The petrology and isotopic signature of the 1530 AD erupted products suggest a zoned magma chamber at a depth of about 6 km periodically fed by basaltic magma from depth (Touboul et al. 2007; Boudon et al. 2008).

Carbon-14 dating of lahar, pyroclastic surge and pumice fall deposits indicate that further activity occurred between 1530 AD and the arrival of European settlers in 1635 AD (Boudon et al. 2008; Legendre 2012). Historical activity has been phreatic, characterized by explosions, episodes of vigorous degassing and ash venting. Major phreatic eruptions occurred in 1797–98 and 1976–77, with minor events in 1690, 1812, 1836–1837 and 1956 (Boudon et al. 1988; Komorowski et al.2005). Hazards associated with phreatic activity include vertical and laterally-directed blasts, ash fall, small-volume non-magmatic pyroclastic density currents, debris flows, acid degassing, potential contamination of the groundwater and aquifer. Major phreatic eruptions could trigger flank collapse and lateral non-magmatic blasts.

The 1975–77 Episode of Eruptive Unrest at La Soufrière

Between 1956 and June 1975 the mean monthly number of recorded volcano-tectonic (VT) earthquakes was 15 and the mean monthly number of felt VT earthquakes was about 0.2 (Dorel and Feuillard 1980; IPGP 1956–2013). This 19-year average is considered the baseline level of VT activity. In contrast to previous phreatic eruptions of La Soufrière (including the last short-lived phreatic eruption of 19–27 October 1956: Jolivet, 1958), a one year period of steadily increasing volcanic seismicity was recorded and felt in Guadeloupe prior to the start of surface activity. This pre-eruptive seismicity was characterized by large numbers of recorded and felt events, and the occurrence of three distinct successive earthquake swarms of increasing released seismic energy. An initial swarm of 30 VTs, one of which was felt, occurred in July 1975. Monthly seismicity continued to increase in the following months and was characterized by two significant swarms of VTs in November to December 1975 (total of 296 VTs and 4 felt events) and in March to June 1976 (total of 2713 VTs and 59 felt events). Although by this stage the rapidly escalating seismicity was about 45 times greater than the baseline monthly rate there were no discernible changes in fumarolic activity (Dorel and Feuillard 1980; Feuillard et al. 1983). However, on 9 June 1976 a minor landslide occurred on the La Ty fault and new fractures were observed on the road at the base of the dome (Feuillard 2011). An explosion occurred at 08:55 local time on 8 July 1976, marking the onset of nine months of complex eruptive activity. The explosion reactivated the 1956 phreatic eruptive fracture and produced 60% by volume (0.6×10^6 m^3) of the total ejected solids from the entire 1976–1977 crisis (Le Guern et al.1980). Table 2 summarises the timeline of key eruptive phenomena, monitoring and phenomenological data (observables) and associated decisions by the scientific team and the public officials.

Table 2: Timeline of the volcanic crisis from July 1975 – June 1977

Date	Eruptive outcome	Key analytical & recorded observables	Key scientific decisions	Key official decisions
24 July 1975 – 7 July 1976: non-eruptive unrest phase: escalating seismicity (3189 recorded VTs, 64 felt VTs)				
31 Jul 75		Seismic swarm 1 24-26/07 (30 VT, 1 felt; 3 stations) 2x 19 year baseline of 15VT/month, 0.2 felt/month	Scientific vigilance – authorities informed	
Nov-Dec 75		Seismic swarm 2: 13/11 to 30/12 (296 VTs, 4 felt) 20x baseline	13/11: MF warns Préfet unrest could lead to eruption	13/11: Préfet requests Volcano Contingency Plan (VCP)
10 Jan 76		Decline in seismic activity to 2.6x baseline followed by increase	19/01: MF 2nd **note to Préfet informing on increasing seismicity (number of VTs and energy)**	10/01: First draft of VCP produced; 4/02: VCP to local authorities for confidential review

24-25 Mar 76		Onset seismic swarm 3: 220 VTs, 12 felt VTs in 24 hr. Total for swarm 3 from 1/03 to 07/07/76: 2713 VTs, 59 felt, 45x baseline		25/03: French government alerted; 1st local press article on unrest - spontaneous evacuation of 200 persons
29-30 Mar 76		March: marked increase in seismicity, 607 VTs, 22 felt, 41x baseline	30/03: HT: 48h visit with GA & GJ: HT gives reassuring assessment, uncertain unrest will lead to eruption, but given an eruption low probability that it will be "cataclysmic"	29/03: Préfet holds crisis response meeting; VCP officially issued defining risk area & evacuation centres; Préfet requests risk map from BRGM
1-12 Apr 76				1-12/04: Préfet reviews VCP with town officials
22-23 Apr 76		120 VTs, 7 felt in 48 hr April: seismicity still increasing, 747 VTs, 14 felt, 50x baseline	20/04: meeting gov Paris, enhance observatory capacity, HT: unrest not exceptional cannot exclude serious outcome, important precautionary measures needed; 23/04: MF reassuring, situation serious but not critical, nothing likely ≤ 4 months given current seismicity trend; 30/04: MF eruption probable July-August if unrest continues	23/04: Préfet press conference, VCP made public; 30/04: Overseas Minister visits Observatory with National Civil Protection to validate VCP, national VCP on file at Paris Civil Protection Office; 22/05: President of France visits, requests zero risk policy
27-28 May 76		100 VTs, 4 felt in 48 hr;		
		May: 611 VTs, 9 felt, 41x baseline		

4-9 Jun 76		4/06: 169 VTs; 8-9/06: 65 VTs 10 felt in 24 h; 9/06: landslide on La Ty fault SE dome fissures on road base dome aligned La Ty fault	02/06: new BRGM risk map sent to Préfet; 14/06: MF note to Préfet, risk map exaggerates risk of PDCs, revision to include ashfall & mudflow risk	
02 Jul 76		June: 668 VTs, 14 felt, 45x baseline	MF 2nd **report to authorities on crisis upon return from Paris**	
8 July – 10 November 1976: Phase 1: 17 explosions, escalating seismicity (11649 VTs recorded, 68 felt VTs), ash & gas venting, mudflows				
08 Jul 76	EX1	8:55 1st and largest EX. Unexpected "blue sky" paroxysm, 48 min tremor, 20 min darkness, new 400 m fracture, cold PDC 1 km, aquifer resurgence, lahar 3.5km, ashfall. Escalating seismicity	MF: reassuring assessment, stronger than 1956, phreatic, no immediate danger	Spontaneous panic evacuation of 25,000, no official order. Traffic regulated. No implementation VCP 1st alert level.
09 Jul 76		Acid vapour jet, pH 1.1-3.2, strong H2S odour to coast, SO2? 609 VT/month, 12 felt VT/month last 3 months (8/04-8/07)		**9-10/07: Préfet evacuates hospitals (1400) and prison. 10/07: Préfet asks population to return to normal life**
13 Jul 76		5 felt VT, 14/07: 14 ppm H_2S ambiant air 3 km SW of crater;		
24/25/27 Jul 76	EX2, EX3, EX4	July: Seismicity escalating 1220 VTs,	13/07: HT+RB+8 scientists arrive+BRGM; HT downplays risk; no magma, precursors to warn of paroxysm; probability of Pelean PDC extremely low, others disagree; 26/07: HT reassures Préfet and leaves for Ecuador, recommends JT to Préfet if activity increases again	22/07: virtually all evacuees back, normal life activities; 25/07: new spontaneous evacuation Matouba (St Claude) 29/07: financial compensation to farmers for ashfall
9 Aug 76	EX5	20 felt, 81x baseline		

12 Aug 76	ash venting	40% fresh vesicular glass ash	**10/08: JT: P(magma present) = 0.25 ; P(large eruption given magma present) = 0.5**	Matouba, Papaye, St Claude evacuated
13 Aug 76	**ash venting**	**1194 VT/month, 17 felt VT/month in 3 months (13/05-13/08) Identification of 40% fresh vesicular glass in ash**	**13/08: RB returns confirms MF + JT, serious escalating unrest, imminent eruption possible, magma at depth**	**17:30 Préfet evacuates ill, elderly, Baillif; 19:30 Préfet evacuates all S Basse-Terre except civil servants. Approx 33600 evacuated since 12/08**
15 Aug 76			MF, RB, JT request urgent meeting of officials "irreversible process, directed PDC impact possible anywhere, risk area must be evacuated"	8:40 Préfet orders phase 2 of VCP & evacuation of 73600 in <24h; remaining 40000 evacuated by 21:00, no injuries. Approx 2000 refused & remain in danger zone
16 Aug 76		725 VTs in 24 h, strongest VT of the crisis, Md = 4.2 (Intensity VI), felt Pointe-**à**-Pitre 60 km away	RB: magmatic gases present, expects EX of 30Mt	Préfet orders 24h evacuation of Volcano Observatory, and states a probability of 1 in 10,000 that nothing happens
21/25 Aug 76	EX6, EX7	24 Aug: 1527 VTs 24 h Strong felt VT, Md = 3.9 (VI) Pointe **à** Pitre		
29 Aug 76		Installation of 2 tiltmeters (USGS 28/08). August: 5989 VTs, ≥ 41 felt, 440x baseline; over last 3 months 2473/month, ≥ 25 felt VT/month (1/06-29/08)	28/08: CA + GA: arrive; 29/08: HT arrives, downplays risk & contradicts RB	
30 Aug 76	EX8 2st **largest**	New fracture splits dome SE, cold PDC 1 km, ballistics & ashfall	EX while scientists (CA, HT, GA...) at summit, HT injured	
31 Aug 76		85% glass magmatic origin in 30/08 explosion tephra Installation of tiltmeters by Los Alamos team (LANL)	HT reassuring, people can return to work during the day without immediate risks; RB + JT leave	

14 Sep 76	EX9	small laterally-directed blast to 1.5 km SE-S, cold PDC 0.5 km, ballistics & ashfall, aquifer resurgence and 8 km lahar	15/09: Official scientific report (GA+CA): evidence precludes any optimistic interpretation of evolving unrest, major events likely in next few months/weeks, dangerous outcome threatening population cannot be excluded; 6/09: official Volcanic risk map (BRGM) issued to Préfet, part evacuated area (Capesterre; Vieux Habitants) now considered much less threatened; 19/09: CA arrives, reports to scientific team absence fresh glass & to Préfet on 20/09; 21/09: CEA report shows no evidence new magmatic gas; 29/09: press conf RF states unrest evolution needs to be monitored carefully	3/09: Local authority CG states situation remains critical, recommends evacuation maintained for 30 days; 6/09: road blocks open 5h30-18h to allow people to return to work in all areas except Basse-Terre, Gourbeyre, St-Claude where individual authorisations are required; 15/09: Préfet designates safe area for day & night time access with new risk map (Capesterre; Vieux Habitants), school restarts in Vieux Habitants, parts of Capesterre; 28/09: installation of siren network where daily access permitted for authorised activities & partial reoccupation
		Mostly H_2S in fumaroles, no or little SO_2, CO_2, low F		
19 Sep 76		Los Alamos reports total absence of "fresh glass" in 8-July-76 ash		
22 Sep 76	EX10	Sept: 1716 VTs, 4 felt, 114x baseline		

2/10/30 Oct 76	EX11, EX12, EX13	No fresh glass in ash, no deformation, tilt not useful	5/10: HT against IPGP's erroneous report fresh glass 8/10: Given decrease unrest & lack magmatic components in erupted ash, MF agrees opening large part safe area (E-SE: Capesterre; Trois Rivières, Gourbeyre, parts of Basse-Terre); 14/10: BM reports lack magma involved, stable gas chemistry and VT hypocentral depths; likelihood rapid onset major eruption decreased, advises continued caution. 26/10: LS favourable reopening Basse-Terre to Baillif area given new analysis and expert opinions since 5 Sept report; 27/10: Scientific Board of IPGP dismisses HT as Head of the Volcano Observatories office of IPGP. HT resigns from IPGP; LS: no evidence magma, no immediate elevated risk	1/10: Day & night access OK as of 4/10 in Capesterre, Vieux Habitants, Trois-Rivières & Gourbeyre, with some areas assessed on individual request; access 5h-18h only in Basse-Terre, St-Claude, parts of Gourbeyre & Baillif;. 8/10: Préfet request to the Minister of the Interior that a international group of experts analyse the situation and formulate recommendations; 11/10: Préfet announces his intention to reopen as of 8 Nov schools in Trois-Rivières and Capesterre and daily activities in Basse-Terre without authorisation; 27/10: Préfet reopens Baillif, Basse-Terre and lower part of Saint-Claude for daily access without authorisation
		12/10: report by Los Alamos shows tilt data is difficult to interpret, confirms lack of magmatic components in the ash, but the increase of sulfur in gases needs to be monitored carefully		
1/6/7/10 Nov 76	EX14, EX15, EX16, EX17	Oct: 2315 VTs, 3 felt, 154x baseline		
11 November 1976 – 4 January 1977: Phase 2: No Explosions, ash & gas venting, lower seismicity (968 recorded VTs, 9 felt VTs), partial reoccupation of evacuated zone				

15-18 Nov 76			17/11: LS declares end crisis, low seismicity & ash emissions; 15-18/11: International Scientific Committee (ISC) validates Aug evacuation, phreatic or deep magma intrusion, estimates low direct risk, recommends return of population	18/11: Overseas secretary declares safe to open all areas including St Claude as of 1 December. Progressive return of ill & administration, Basse-Terre schools to reopen 2/1/77 – only 7000 evacuees left in relocation centres
30 Nov 76		Nov: 1040 VTs, 0 felt, 69x baseline; 1843 VT/ month last 3 months, 0 felt (30/08-30/11)		Population from St Claude & Basse-Terre unwilling to return given frequent ashfalls & degassing
1-15 Dec 76		Strong gas jet at summit, gas temperature 184°C, ejection blocks; Dec: 399 VTs, 6 felt, 27x baseline		**15/12: End of evacuation, partial reoccupation**
5 January – 1 March 1977: Phase 3: 9 explosions, ash & gas venting, declining seismicity (475 recorded VTs, 7 felt VTs), total reoccupation of evacuated zone				
5/13/14 Jan	EX18, EX19, EX20	Frequent ashfalls on upper Saint-Claude, lapilli size ballistics to 1.3 km SW on 29/01; strong roaring; gas (H_2S; SO_2) odour; Jan: 312 VTs, 4 felt, 21x baseline		29/01:Matouba & Papaye villages closest to volcano to SW isolated overnight due to ash on road, people worried
15/17/29 Jan	EX21, EX22, EX23			
13/15 Feb	EX24, EX25	Ashfall on upper Saint-Claude; Feb: 179 VTs, 3 felt, 12x baseline		
01-Mar 77	**EX 26: Last & 3rd largest**	Ashfall on upper Saint-Claude, Basse-Terre, and Baillif; 4/03: potable water contamination by ash. SO_4^{-2} > maximum accepted; F^- almost to maximum accepted. **297 VT/month (20x baseline) 4.3 felt VT/month (1/12-28/02)**		
2 March – 15 June 1977: Post-eruptive unrest phase: no explosions, no ash venting, declining seismicity (212 recorded VTs, 5 felt VTs)				

28-Mar-77		Reported stinging in nose from SO_2 at base of dome; March: 153 VTs, 4 felt, 10x baseline		
15-Jun-77		April: 32 VTs, 0 felt; May: 19 VTs, 1 felt; June: 15 VTs, 0 felt, pre-crisis baseline level. Sustained degassing at summit & periphery		Eruption officially ends, clinics and hospitals reopen

Dates in bold are pivotal dates when probabilities of eruptive outcome was queried using the BBN (Figure 5). Seismic data within each phase of the eruption is discontinuous and given only on specific key dates, the total seismicity is given in the heading of each phase. Detailed data is available in references below. EX: phreatic explosion; PDC: pyroclastic density current; RB: R. Brousse, Paris based, Expert 1 Head Government appointed Scient. Com.; HT: H. Tazieff, Paris based, Head IPGP Volc Obs, Expert 2 requested by Préfet; MF: M. Feuillard, IPGP, and Head Soufrière Volc. Obs., JT: J Tomblin, Head SRU Trinidad, arrives 3/08/76 invited by Préfet (intern. expert suggested by HT on 29/07/76 given HT's departure to Ecuador); GJ: G. Jobert; Dir. IPGP from 01/08/71 till 1/08/76; CA: C. Allègre; Dir. IPGP as of 08/08/76; BRGM: Bureau de Recherches Géologiques et Minières; GA: G. Aubert, Assistant Dir. INAG, Nat. Institute of Astronomy & Geophysics (CNRS); RF: Richard Fiske, expert from USGS; LS: L. Steinmetz, seismologist, head of IPGP scientific team in Guadeloupe from 29/09/76 to 02/12/76; CG: Conseil General Guadeloupe, local government; International Scientific Commission (ISC): Chair: F. Press; members: S. Aramaki, F. Barberi, J. Coulomb, R. Fiske, P. Gasparini, C. Guillemin, G. Sigvaldason); BM: B. Minster, head of IPGP scientific team in Guadeloupe from 03/09/76 to 17/09/76; PDC: pyroclastic density current (nuée ardente). Data compiled from: Dorel and Feuillard, 1980; Feuillard et al. 1983; Préf. Guadeloupe 1977; De Vanssay, 1979; Lepointe, 1999; Leguern et al. 1980; IPGP, 1956–2013; Smithsonian Inst. 1976, 1977; Sheridan, 1980; Feuillard, 2011; Boudon et al. 1988; Loubat and Pistolesi-Lafont 1977; Komorowski et al. 2005; Beauducel,2006.

Hincks *et al*

Hincks *et al. Journal of Applied Volcanology* 2014 **3**:3 doi: 10.1186/2191-5040-3-3

The 1975–1977 unrest and eruptive crisis lasted 22.5 months from 24 July 1975 until 15 June 1977, during which 16493 VTs were recorded of which 153 were felt (Dorel and Feuillard 1980; Feuillard et al. 1983; IPGP, 1956–2013; Feuillard 2011) A total of 26 explosions occurred in two phases over the 9 months from 8 July 1976–1 March 1977. The eruption can be divided into five main phases: a) a pre-eruptive unrest phase from 24 July 1975–7 July 1976 with steadily increasing and escalating seismicity (3189 VTs, 64 felt VTs); b) eruptive phase 1 from 8 July to 10 November 1976, the most intense in terms of explosive activity (17 explosions) and seismicity (11,649 VTs, 68 felt VTs); c) eruptive phase 2 from 11 November 1976 to 4 January 1977 (the least intense) with no explosions, frequent ash venting episodes, and a decrease in recorded seismicity (968 VTs; 9 felt VTs); d) eruptive phase 3 from 5 January to 1 March 1977, which showed a renewal of explosive activity (9 explosions, some amongst the largest of the entire crisis) and a decline in seismicity (475 VTs, 7 felt VTs); e) a post-eruptive unrest phase beginning 2 March during which seismicity returned to almost background levels by 15 June 1977, formally marking the end of the eruption. Excluding volcanic tremor, recorded seismicity was poorly correlated with eruptive phenomena.

The 1975–77 eruption produced an estimated 10^6 m^3 of non-juvenile tephra deposited as very fine, thin ash up to 10–15 km west and south of the summit (Heiken et al. 1980; Le Guern et al. 1980). Ballistic blocks (a few kilogrammes to several tonnes) were ejected up to 1.6 km from the vent. The eruption was accompanied by morphological changes including opening of two new major sets of fractures (several hundred meters long, several decameters wide) in the dome, and the widening and deepening of historically old craters and fractures. Small-volume pyroclastic density currents from the explosive vents extended to about 1 km from the dome (Sheridan, 1980) before transforming into debris flows that formed from the resurgence of perched aquifers and reached up to 3.5 km to the east and south. The eruption was accompanied by low-temperature (100-200° C) degassing of H_2O and minor quantities of CO_2, H_2S, SO_2, as well as acid condensates (HCl, HF, Br) and a sustained renewal of fumarolic activity on and at the periphery of the dome.

Phreatic tephra consisted essentially of old hydrothermally-altered material from the dome and old pyroclastic fragments from the nearby Echelle scoria cone. Several authors (Marinelli 1976; Brousse et al. 1977; Heiken et al. 1980) reported that the phreatic products contained up to 10 % by weight of fresh unaltered vitreous andesitic fragments - these were later identified as material from the underlying 1530 AD pumice fallout deposits, although this was not understood at the time. This led to major scientific controversy as to whether there was evidence of juvenile material and therefore fresh magma in the conduit, compatible with the rapidly escalating and publicly obvious unrest. In the absence of conclusive monitoring data and independent evidence of a magmatic component, this single observation had a pivotal influence on the scientific management of the crisis, justifying a precautionary response by the authorities.

Two models have been proposed to explain the 1975–77 eruption. The more widely-accepted model interprets the crisis as a still-born magmatic event (Feuillard et al. 1983). Seismic activity from a depth of about 6 km is compatible with the inferred depth of a magma reservoir, and observations of SO_2 provided evidence for magmatic unrest. Analysis of Cl levels in thermal springs sampled in the 15 years following the crisis has attributed the cause to shallow magma intrusion (Villemant et al.2005; Boichu et al. 2011), and the chlorine isotope signature of thermal spring waters is markedly magmatic (Li et al. 2012). Feuillard et al. (1983) proposed that abnormal heat flux, either caused by magma differentiation of the 1530 AD magma batch or magma chamber replenishment, led to formation and propagation of fractures towards the surface, enabling migration of magmatic gases into the hydrothermal system. The hydrothermal system acted as a buffer, dissipating heat and magmatic gases and inhibiting magmatic eruption. Recent modelling of the flux of noble gas isotopes ($^4He/^3He$; R/Ra) in hydrothermal fluids suggests that either new magma is being continuously injected in the 6 km deep magma chamber, with a fresh batch of magma emplaced between 1959 and 1962 (Ruzié et al. 2012) or that magmatic gases are being transferred into and through the shallow magma chamber. The attendant increase of heat flux and migration of fluids into the locally sealed hydrothermal system led to progressive pressurization and the phreatic activity of 1976–77.

A second model, proposed by Zlotnicki et al. (1992), does not invoke physicochemical changes in the magma reservoir. Rather, it proposes that aquifers become isolated (sealed) by structural readjustments or deposition of impermeable clay minerals from hydrothermal activity. This sealing is assumed to limit convective heat transfer from depth, leading to pressurization. Phreatic eruption occurs when overpressures are sufficient to cause fracturing.

The 1976–1977 eruption engendered significant and recurrent disruption and risk to the population (Le Guern et al. 1980) largely due to: 1) frequently and strongly felt volcanic seismicity; 2) atmospheric contamination by acid gases (H_2S, SO_2) and fine corrosive volcanic dust rich in acid condensates and Ca-sulfate that sometimes also contained non-negligible quantities of silica polymorphs; 3) contamination of potable spring waters and water tanks due to soluble acid condensates (including halogens such as fluorine, chlorine, bromine) and other trace elements adsorbed on the surface of the erupted ash; 4) chemical and mechanical consequences of the contamination of crops and grazing land due to acid condensates and trace elements adsorbed on the ash, in particular fluorine.

The 6-month evacuation of around 73,000 people (with an estimated 2000 remaining) caused severe socio-economic difficulties for the population in southern Basse-Terre and the island as a whole, and had a profound and long-lasting influence on society. The cost of evacuation has been estimated at 60 % of the total annual per capita Gross Domestic Product (GDP) of Guadeloupe in 1976 (Lepointe,1999; Blérald, 1986) or about 342 million USD using 1976 currency rates (Kokelaar 2002); this excludes the losses of uninsured and other personal assets including open-grazing livestock and farm animals. Hence, the crisis ranks amongst one of the most costly of the 20th century (Annen and Wagner 2003), although there was no loss of life.

A Bayesian Belief Network for Guadeloupe 1975–77

A BBN has been developed to describe the fundamental processes and interactions governing volcanic activity and the state of unrest at La Soufrière from 1975 to 1976 (Figure 3). This is an extension of the model developed in Hincks (2007), following re-analysis of the literature and reports about the 1976 crisis as part of the European Union EXPLORIS project (Explosive Eruption Risk and Decision Support for EU Populations Threatened by Volcanoes, Komorowski et al. 2004) and the 2009 Agence Nationale de la Recherche funded CASAVA project (Compréhension et Analyse des Scénarios, Aléas, et risques Volcaniques aux Antilles). This BBN is a simplified conceptual model of the volcanological system depicted in Figure 2, informed by understanding of the causal relationships between processes at depth and resulting surface manifestations of activity. As the aim here is to perform a retrospective analysis of the 1976 crisis, the model is based on what was then contemporary knowledge of the volcanic system and includes only observations available at the time. The basic structure could easily be adapted to build a model for renewed activity, incorporating additional nodes and arcs to reflect the increased scope, resolution and frequency of modern monitoring, and current understanding of the various interactions and time-scales involved (e.g. rates of degassing), informed by both observation and numerical simulation.

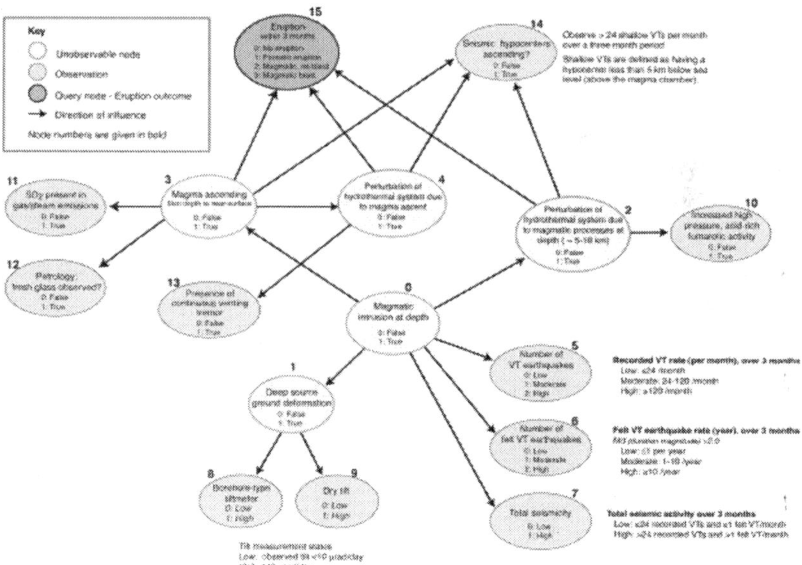

Figure 3: Bayesian belief network diagram for La Soufrière. BBN showing the relationship between volcanic processes, states and observations available in

1976, used to infer future activity. Nodes represent both hidden (grey) and observable (blue) states. The arcs between nodes represent conditional dependencies (e.g. direct causal relationships or influence) and are characterized by conditional probability tables (CPTs). The arrows indicate the direction of influence e.g. venting tremor is believed to be a sign of perturbation of the hydrothermal system resulting from magma ascent. In this case, all conditional probability distributions (and associated uncertainties) were obtained by expert elicitation (for example, the probability of observing SO_2 given that magma is ascending), the network structure being agreed by the group prior to elicitation. For data rich applications parameters can be estimated purely from data, or a combination of observations and prior knowledge (e.g. using Dirichlet priors or other parameter constraints: Niculescu et al. (2006). The basic difference between this and a logic or event tree representation is explained in the Methods section and Figure 4.

All nodes are discrete, with mutually exclusive and exhaustive states (Figure 3). Arcs between nodes indicate direct conditional dependencies (represented by conditional probability tables - CPTs), with the arrow showing the direction of influence. Observations (nodes shown in blue) can then be used to make inferences about unobservable or 'hidden' states of the system (grey) and about the outcome of interest - whether or not an eruption will occur (the query node). This is a static network which, unlike an event tree, does not explicitly model time dependency, i.e. variables are evaluated only at a particular point in time. The probability of eruption and other unobservable states can, however, are evaluated at discrete time steps to give a changing 'hazard forecast'. Temporal associations between nodes can be incorporated in a dynamic BBN - see Future Developments.

In order to enumerate network parameters (conditional probabilities) and associated uncertainties for this analysis, we conducted a structured expert judgement elicitation over two days in November 2007 (see Methods, below). The group (the authors and colleagues, see acknowledgements) comprised seven volcanologists with expertise covering a range of relevant disciplines, some of whom were very familiar with the 1976 events. With the Cooke approach, a group of seven experts would be regarded as close to a minimum quorum, but here it is considered sufficient for deriving quantitative values for the BBN. Following an initial discussion of the network structure and related elicitation questions, the experts were asked to provide their personal 5, 50 and 95 percentile estimates of different volcano state probability values, taking into account uncertainties associated with the various processes, data and interpretations of evidence. After pooling these opinions into joint, group distributions (see Methods), we use the results to populate conditional probability tables (CPTs) for each node in the BBN. For example, one such question for node 1 was phrased (using the present tense) as follows: "What is the probability of ground deformation occurring due to a deep source in La Soufrière region, given magmatic intrusion has occurred at depth (5–10 km)?". The elicitation questionnaire and results are provided as supplementary material to this paper With respect to the elicitation, a key feature of the procedure is that it encourages experts to state, independently after group discussion, their true opinion; this limits direct peer influences and other biases (Aspinall 2006, p28).

To frame the elicitation questions we chose a time scale of three months to evaluate the likelihood of observing any given phenomena or sustained level of activity. Recognizing that uncertainty in hazards forecasts increases the further ahead one looks, this time period is a pragmatic choice, appropriate in terms of decision-making in a volcanic emergency. Moreover, during the pre-eruption unrest phase at La Soufrière from June 1975 to July 1976, earthquake swarms that lasted about 1 month were separated by periods of lower seismic activity on the order of 2–3 months (Dorel and Feuillard 1980; Feuillard et al. 1983). At the Soufrière Hills Volcano, Montserrat, a time scale of six months was chosen for hazard and risk assessment during the volcanic emergency (Sparks and Aspinall 2004), this period being long enough for authorities to make and implement mitigation plans but judged short enough by the science assessment team for making useful hazards forecasts. In practice, the forecast time period can be determined by the situation and other demands, with the network nodes and structure designed to accommodate multiple time periods if necessary.

Although observations available in 1975–77 only are used in our analysis, knowledge of volcanic processes and the assessment and interpretation of monitoring data have advanced since then. The group felt it was impossible to transport themselves back fully into the state of general volcanological knowledge and mindsets of the time so the elicitation process inevitably incorporates some elements of contextual advances in the science.

METHODS

Bayesian Networks

The basic concept behind the treatment of uncertainty in Bayesian Belief Networks is conditional probability (Jensen and Graven-Nielsen 2007). A BBN is a directed acyclic graph and comprises a set of variables (nodes) with either continuous or discrete state-sets, together with a set of directed links or arcs representing the direction of causality or influence between the nodes (see basic example in Figure 4a). Nodes can represent observable or hidden states of the system. A link between two nodes is described by a conditional probability distribution (CPD), with multiple state distributions entered on a Conditional Probability Table (CPT). There are similarities with the logic or decision tree (a BBN can have a tree structure, or be approximated by a tree); however, the BBN captures conditional independence and in some cases can represent the system more compactly. The tree (Figure 4b) has a hierarchical as opposed to a network structure, and represents a sequence of events occurring in time culminating in the outcome of interest. In a BBN, decision or logic tree, node states are mutually exclusive and exhaustive, but this is not necessarily the case in an event tree (e.g. see Marzocchi et al. 2006).

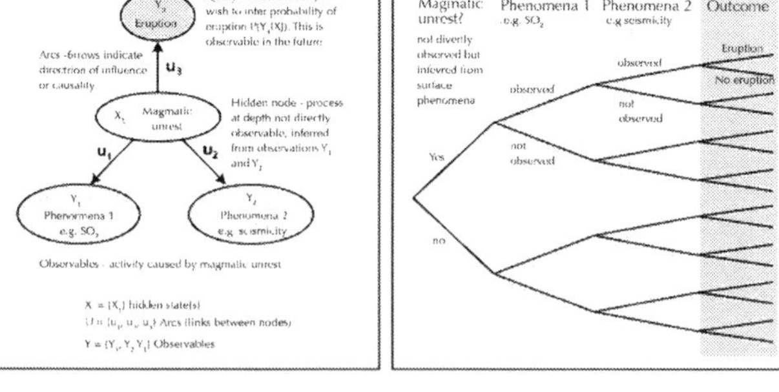

Figure 4: Comparing a Bayesian Network and logic or event tree. (a)This is an elementary BBN representing a very simple volcanological model to infer the probability of eruption resulting from magmatic unrest. Unrest (shaded grey) cannot be observed directly, and must be inferred from observations (e.g.

seismicity and SO_2 emission, shaded blue). The query node (shaded pink) is the outcome of interest - eruption. For a generic Bayesian Network B = {X, Y, U}; X represents the set of unobserved or hidden states, Y the set of observable states, and U the set of arcs or directed links between nodes. Arrows indicate direction of causality or influence. For the simple case where nodes are assigned discrete states, node relationships are described by conditional probability tables (CPTs). States must be mutually exclusive and exhaustive. (b) Presents an equivalent logic tree representation of the basic BBN in (a). For simplicity two alternative states are shown for each node.

In this study, OpenPNL[b] (an open source C++ library initially developed by Intel) was used for the computational BBN. A separate "link connection strength" package, an add-on to OpenPNL by Ebert-Uphoff (2007), was used to calculate mutual information (described below). There are numerous alternative software tools with graphical user interfaces which make BBN inference and analysis accessible. Many also provide API (Application Programming Interface) functionality enabling custom code to be written for automated inference and data input/output. Examples include GeNIe and SMILE (Decision Systems Laboratory 2013), Netica (Norsys 2013) and Uninet (Cooke et al. 2007).

[b]OpenPNL is available at http://sourceforge.net/projects/openpnl/ *webcite* andhttps://github.com/crishoj/OpenPNL

In the static BBN presented here, data from the 1976 episode were limited and it was necessary to use expert judgement to fully describe the stochastic and scientific uncertainties. The ability of the volcanologists to quantify uncertainty was therefore critical to the reliability of the model, and we used a structured elicitation procedure (described below) to obtain collective, quantified uncertainty distributions.

BBN Network Structure

The network structure and node states for the Guadeloupe BBN are shown in Figure 3. There are three different node types: 'hidden' or latent state nodes; 'observable' nodes, and the all-important eruption outcome 'query' node. The state of the hidden nodes (unobservable volcanic processes) and eruptive outcome can then be estimated from the input observations using Bayesian inference (see Aspinall et al. 2003). The following section describes each of the nodes in more detail.

Documents the elicitation questions used to populate the conditional probability tables for each node in the BBN.

Hidden Nodes

These represent unobservable volcanic states and processes that can only be inferred from observations.

Magmatic Intrusions at Depth

Depth is defined as 5 km or greater, in the region of the magma chamber. Node 0 has no parents, therefore experts were asked to estimate 5, 50 and 95 percentile values to characterize a probability distribution that new magma was being intruded in the *absence*of any volcanological evidence – i.e. the baseline probability of intrusion taking place at La Soufrière at any point in time (see– elicitation questions). This is a simple binary node – the state is either true or false.

Deep Source Ground Deformation

This is defined as a source at 5 km or greater depth, giving rise to wide field surface deformation. It is anticipated that only magmatic processes could result in such deformation. Node states: true/false.

Perturbation of the Hydrothermal System Due to Magmatic Processes at Depth

The network considers the effects on the hydrothermal system due to magmatic processes at depth (typically 5–10 km), and magma ascent (above 5 km depth, node 4 below) separately. Identification of either effect can be complicated by exogenic forcing (tectonic and meteorological). Node states: true/false.

Magma Ascent

Ascent is defined as any movement of magma above the magma chamber (around 5 km depth) toward the near-surface. Magma need not necessarily reach the surface and could stall at shallow depth. This network node represents the probability of ascent over a three month period, given magmatic intrusion is occurring at depth. If there is no magma intrusion then the probability of hydrothermal perturbations due to magmatic processes at depth is assumed to be zero (and therefore not elicited). States: true/false.

Perturbation of the Hydrothermal System Due to Magma Ascent

As for node 2, effects can be complicated by exogenic forcing. If there is no magma ascent then the probability of hydrothermal perturbations due to magma ascent is zero (therefore not elicited). States: true/false.

Observable Nodes

Figure 5 (lower panel) shows the time series of observational evidence input to the BBN, replicated from 1976 records and narrative accounts from Guadeloupe. As this is intended as an indicative example rather than a complete chronology of observations, these inputs span the whole crisis but focus on key dates when significant observations were made, according to contemporary reports. Computation time is very short, a few seconds, so a modern day BBN could be updated frequently and coupled with real time monitoring data streams to give near continuous hazard estimates. The following summary describes the chosen thresholds for different levels of activity for each of the observables, with some (modern) commentary on the available data, processes and reliability of evidence. The commentary reflects discussions and decisions among the expert group, referenced where possible. Volcano-tectonic earthquakes are denoted as VT.

Number of VT Earthquakes

The baseline rate for Guadeloupe was assumed to be approximately 12 VT earthquakes per month (Dorel and Feuillard, 1980; Feuillard et al. 1983). Recent access to more detailed reports (IPGP, 1956–2013) gives a revised baseline rate of 15 recorded VTs/month and 0.2 felt VTs/month (as quoted in the description of eruptive unrest earlier in the manuscript); this, however, has little effect on the original findings.

Three states are defined: low (≤ 24/month, up to double the baseline rate), moderate (between 24 and 120/month) and high (greater than 120/month, or in excess of 10 times the baseline).

Number of Felt VT Earthquakes

Felt earthquakes are typically VT events with a duration magnitude (Md) of approximately 2 or greater. This node has three states: low (≤ 1 felt earthquake per year), moderate (1-10/year) and high (>10/year). The baseline rate from historic records for the period from 1956 through 1975 assessed in contemporary reports (Smithsonian Institution1976; 1977) was assumed to be approximately 0.6/year (although note that as for node 5, IPGP (1956–2013) gives a revised baseline of 0.2 felt VTs/month). There are good data on felt earthquakes associated with previous episodes of unrest (Jolivet 1958; Dorel and Feuillard 1980), and although various factors influence detection and reporting (population distribution, survival of records etc.) this non-instrumental observable is assumed to be an informative parameter, with a relatively consistent historical detection threshold. Felt VTs can also be used as a proxy for seismic energy release. However, felt VTs could be non-magmatic in origin, e.g. due to activation of faults related to disturbance of the hydrothermal system. It is possible that magmatic intrusion may occur but not be accompanied by detected or felt VTs[c].

[c]As, for example, in two eruptions of the nearby Soufriere of St. Vincent volcano in 1971 and 1979 (Aspinall et al. 1973; Shepherd et al. 1979). In these instances, there was almost no (recorded) precursory seismicity: in 1971 the nearest seismometer on St Vincent was 30 km from the volcano, while in 1979 there were seismometers on the volcano but only low-level instrumental tremor was detected and then only in the 12 hours before the first explosion.

Total Seismicity

This node represents the observation of deep seismicity, and the likelihood of it being caused by magmatic intrusion. A high level of activity is defined as more than 24 recorded VT signals and more than 1 felt earthquake per month. Non-magmatic triggers could include tectonic activity associated with the regional fault system and high fluid pressures and stresses in the hydrothermal system.

Borehole-type tiltmeter observations

Following vigorous ash venting on the 30 August 1976, scientists from the Los Alamos National Laboratory deployed four novel borehole-type tiltmeters on the southwest flank of La Soufrière, at distances of 0.8 to 8 km from the vent in temporary surface installations. Measurements for the 22–23 September, 27 September and 2 October 1976 are given in West et al. (1976). In 1976 this was new, unproven, experimental technology. The instruments were not fixed in boreholes and hence the data are considered unreliable, with non-volcanic causes of tilt (e.g. ground surface instability, wind, heavy rainfall). For this node, a high tilt rate is defined as >10 μrad/day.

Dry Tilt Data

In 1976 dry tilt was considered to be an accurate and tested method, likely to detect widespread deformation. In August 1976, R. Fiske and K. Kinoshita set up four dry tilt stations on the SW flank of La Soufrière. Data for the period 29 August - 30 September are presented in Smithsonian Institution (1976) reports. The report states that tilt measurements from 1–16 October lacked coherence, likely due to effects of rainfall and evaporation at the stations. Smithsonian Institution (1977) reports "ground deformation measurements ... (pendulum, borehole, dry tilt) have indicated no significant changes in shape". We therefore assume no evidence of deformation from dry or borehole-type tilt for 30/11/76 and 1/3/77. High dry tilt is defined as >10 μrad/day.

Increased or Decreased Pressure, Acid-rich Fumarolic Activity

Magmatic processes (degassing, interaction with groundwater etc.) can perturb the hydrothermal system and increase fumarolic activity. The presence of highly water-soluble halogen species (e.g. acids such as HCl, HBr) in thermal springs can be explained by magma intrusion and degassing that interacts with the hydrothermal system (Boichu et al. 2011estimated a volume of 0.01 - 0.52 km^3 of intruded magma). Halogen acids are stable in water and largely unaffected by cooling and decompression (Villemant et al. 2005). Fumarolic activity, Cl content of thermal springs and seismicity slowly declined in the 10–15 years following the crisis. The slow decline could be associated with reducing supply of magmatic fluids. Alternatively changes in porosity and permeability, along with self-sealing of the host rock and conduit, could reduce rates of surface degassing without the deep source flux of magmatic gases necessarily changing. The hydrothermal system could also be affected by tectonic and weather-related processes, although this was considered by the expert group to be low probability.

Temperatures of fumaroles and hot springs had been measured since 1956 (Zlotnicki et al. 1992). Limited gas chromatography data existed, but problems with sampling (reaction of gas species in the sample vial during and after sampling) make these unreliable measurements (Feuillard et al.1983). Given this uncertainty, and that regular sampling of fumaroles did not begin until after the crisis in 1979, this node has simply two states (true/false) to capture any reported increase in fumarolic activity in the absence of reliable geochemical data. Between 1970 and 1976 the only active fumaroles were at the base of the dome, and in 1976 fumarolic activity developed rapidly on new and re-opened fractures on the summit, flanks and base of the dome. Heiken et al. (1980) report near-continuous background fumarolic activity during the period August-October 1976. Activity decreased rapidly towards the end of the crisis, ceasing first at the summit, then at the periphery in 1977, with the north fault fumarole disappearing in May 1977 (Zlotnicki et al. 1992). For purposes of this analysis we therefore assume elevated fumarolic activity was present from mid-1976, returning to a low level by 30 November 1976.

SO_2 Present in Gas/Steam Emissions

This node represents the probability of detecting SO_2, given magma ascent (or otherwise). SO_2 is regarded as diagnostic of shallow depth magmatic origin (Villemant et al. 2005), however, chemical and other processes (decompression, cooling, distance from source, chemical reactions etc.) can affect concentration, speciation and the time lag from release to detection.

In assessing SO_2 there is needed to account for false positive and false negative observations. For example, the potential for SO_2 emissions due to deep magmatic unrest, but no magma ascent above 5 km the meteoric system can also act as a filter or sink for chemical species that might otherwise signify magmatic activity (e.g. scrubbing of SO_2 or halogen species). La Soufrière has an extensive hydrothermal system (Villemant et al. 2005; Zlotnicki et al. 2006) and scrubbing due to tropical rainfall will increase the chance of a false negative observation. Conditions at the time adversely affected the quality and frequency of observations, and as a result there are no published SO_2 data from the time of the crisis (only pH). The pH is not diagnostic if anions (SO_4^{2-} and SO_3^{2}) are not analysed to rule out the contribution of acid halogens (e.g. HCl) to the acidity of fumarolic condensates. Moreover false positive secondary SO_2 can be formed at fumaroles due to rapid oxidation of H_2S by the atmosphere or bacteria.

Petrological Evidence - Observation of Juvenile Material

This is defined as observation of abundant, unaltered fresh juvenile glass. On 12 August 1976 some scientists identified pumice in ejecta from steam explosions, and this was presented as strong evidence for fresh juvenile material. However, this sample may have been recycled material from the previous magmatic eruption in 1530 AD or even misidentified clay minerals. At the time juvenile glassy material was typically assumed to be vesicular, but recent research suggests otherwise. For example, non-vesicular, microlite-rich glassy fragments of the Mt St Helens cryptodome have been identified (post hoc) by Cashman and Hoblitt (2004) as a magmatic precursor. Here the experts were required to evaluate the probability of a false negative result or misclassification. Fresh glass may be present but not identified in the sample, magma may be ascending but fresh material may not be present at the surface, and old products remobilised by phreatic activity could be misidentified as fresh material.

Presence of Continuous Venting Tremor

Continuous tremor can be generated by gas escape or fluid flow, and episodes of tremor can be associated with phreatic explosions (e.g. Barberi ET al.1992; Young et al. 1998; Nakada et al. 1999). It is possible (although much less likely) that non-magmatic processes such as changes in groundwater circulation or atmospheric pressure could perturb a hydrothermal system and generate low-level seismic tremor (e.g. Aspinall et al. 1976).

Seismic Hypocenters Ascending

This is defined as the emergence of an average of 24 or more VT earthquakes above 5 km depth (above the magma chamber) over three months. Such events could be caused by magma ascent (movement of magma, rock fracturing, and gas escape) or alternatively hydrothermal processes driving movement of fluids, gas or steam, or fracturing. By considering a 3-month period of activity significantly above the baseline, the aim is to make it easier to differentiate between ascending magmatic and static non-magmatic drivers, as it was considered unlikely that shallow seismicity would persist at such elevated and escalating levels without an element of upward-moving magmatic activity. Discussion of Hirn and Michel's (1979) post-crisis analysis of seismic hypocentres during the elicitation may have influenced the experts here – highlighting the difficulty of performing a reanalysis some 40 years on.

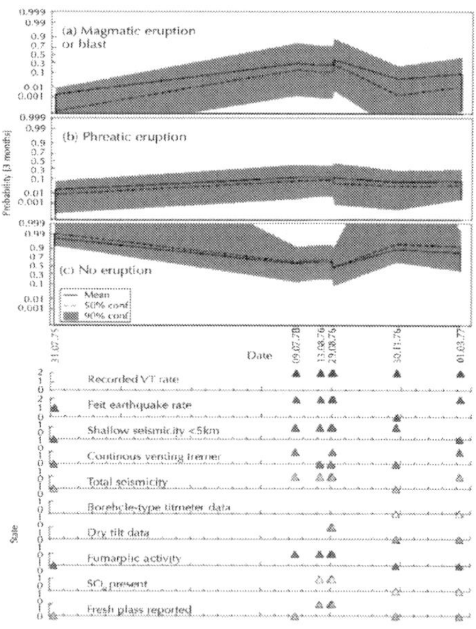

Figure 5: La Soufrière Volcano observations and eruption probabilities for July 1975 to March 1977. Upper three panels: time variations in BBN prob-

ability estimates for: (a) a magmatic eruption or magmatic blast; (b) a phreatic eruption, or (c) no eruption, given observation states shown in the lower part of the figure. The unbroken black line denotes the expected (mean) probability estimate and the dashed line the median, as determined by Monte Carlo re-sampling of BBN input distributions; the shaded bands show the corresponding 5–95 percentile ranges, indicating the uncertainty in the forecast probability. Lower ten panels: sequence of observation states input to the BBN to estimate probability of eruption (see text).

Query Node

Eruption within 3 Months

The outcome (in any subsequent three-month period) is defined as the probability of: (a) no eruption; (b) a phreatic eruption; (c) a magmatic eruption, or (d) a magmatic blast.

Expert Elicitation

A structured expert elicitation is a formalized method for assimilating group judgements in a robust and reproducible way. In a crisis, a carefully targeted elicitation can be completed in an hour or so, including expert calibration, and the findings processed within a further hour, as happened many times in Montserrat in 1995–96 (e.g. Aspinall et al. 2002, pp82-83).

In the present exercise, individual expert uncertainty judgements were combined with equal weights using the "Classical Model" formulation (Cooke 1991; Aspinall and Cooke, 2013) and its implementation in the EXCALIBUR software package (Cooke et al. 2000). The elicitation questions [see 1] were structured to obtain enough information to enumerate all CPTs, and hence fully characterize the BBN. To quantify uncertainty in each parameter in an elemental distribution form, the volcanologists were asked to provide lower and upper tail quantile markers, corresponding to 5% percentile and 95% percentile values, together with a median estimate to locate central tendency. Because of the small size of the group and the variety of specialisms involved, the Classical Model performance-based differential weighting option was considered not to be appropriate; instead, equal weights combinations of experts' uncertainty distributions were computed. For each variable, this took the form of a 'joint' estimate of the relevant 5, 50 and 95 percentile values, expressing the spread of uncertainty for each BBN node item; these quantiles were used to fit and parameterize standard statistical distributions[d] (see section below "BBN parameterization and evaluation").

[d]Both Dirichlet and Generalised Trapezoidal distributions were compared, the Dirichlet distribution gave a better fit to the joint DM quantiles and was used in the final analysis.

Elicitation Responses

The outcomes of the elicitation exercise are provided in an. Most of the responses to questions demonstrate clearly (even four decades after the event) that considerable uncertainty would still attend nearly every aspect involved in assessing the internal state of La Soufrière volcano from observations, and that the diagnostic power , in terms of eruption forecasting, is generally weak. This would have been even more so in 1976. However, this approach enables expressions of scientific uncertainty in more objective terms than possible at the time.

By and large, elicitation responses are reasonably coherent across the group, but the preponderance of wide credible intervals tends to mask any systematic differences between experts. Some of the results are informative with regard to evidential value of particular types of observations. Here we consider three specific target items (out of the 62 elicited, in total) to illustrate some generic issues.

Figure 6 shows the results of the elicitation relating to a question on the probability of observing/detecting ground deformation from a deep source in La Soufrière region. Two versions of this question, involving different conditional stipulations, were elicited: the first given there is *no*magmatic intrusion/unrest at depth, and the second given there *is* magmatic intrusion or unrest at depth. In this case, most experts judged that widespread ground deformation would accompany either intrusion or magmatic unrest. Five experts showed a high degree of confidence in the positive link, whereas the other two experts, whilst agreeing that this link was very likely, also gave a very wide uncertainty range. In contrast, all but one expert placed the likelihood of having either intrusion or magmatic unrest but no surface deformation as low. The same two experts again gave much wider uncertainties to this question than their other colleagues.

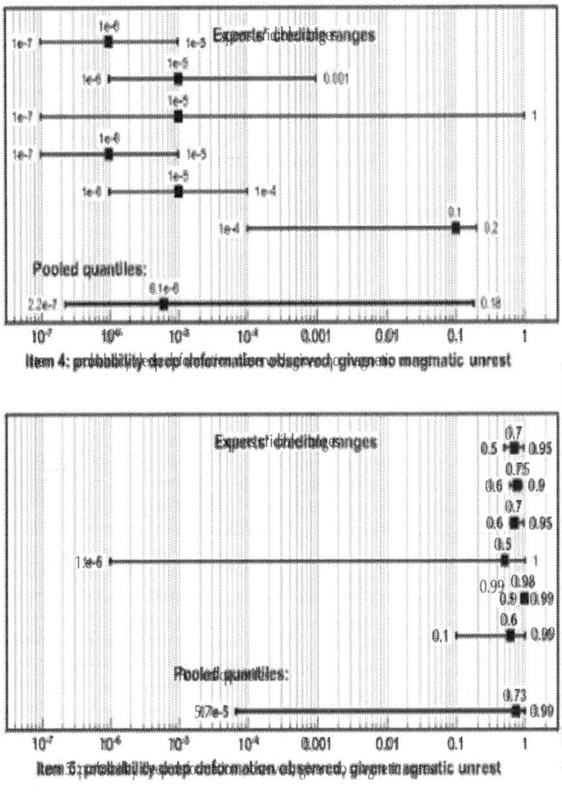

Figure 6: Example expert range graph Plot showing the individual experts responses to a question on the probability of observing/detecting ground deformation from a deep source in La Soufrière region.

These results reflect the widespread view that surface deformation can be equated with intrusion of new magma, magma chamber replenishment or some internal magmatic event, such as convective over-turn, any one of which processes will increase pressure. The minority two experts felt that it was possible to envisage a deep magma system perturbed in some way that would not be necessarily accompanied by surface deformation. In an ideal exercise, these findings would be debated and all experts then re-elicited, having discussed the reasons why colleagues came to different conclusions. The principle here is not to enforce consensus but to be sure that all the pertinent evidence and arguments have been presented, and equally well-understood by the group.

Target Item 11 (Q12 on questionnaire) concerns false positives for SO_2detection. Some colleagues considered that there is no diagnostic power in this information, but others suggest there is a small chance of it having evidential worth. The 1976 crisis occurred before the availability of high quality SO_2 measurements from ground instruments and from satellite remote sensing. SO_2 observations (now routine) would likely now be given more weight However, experience in Montserrat indicates that SO_2 data commonly remain enigmatic or ambiguous; for example, a decrease in SO_2 might mean either a decline in deep activity and hence a diminution of eruption likelihood, or that gas has become trapped enhancing the prospect of an imminent explosion. As with most observables, SO_2 has more evidential worth when analyzed in conjunction with changes in other observables.

With Target Item 36 (Q26 row 4 on table,), four experts seem convinced that there is a high probability of seeing shallow VTs when the hydrothermal system is perturbed by deep magma action but magma is not ascending, while the other three are much more cautious and gave far lower probability values. With respect to Target Items 51 and 52 (Q27 two rows on table,), two Experts (2 and 3) diverge from colleagues in estimating 'No eruption' or 'Phreatic eruption' probabilities when the hydrothermal system is perturbed by deep magmatic action but magma is not ascending. These are examples of "two schools of thought", where further discussion might either resolve the issue to everyone's satisfaction (i.e. some consensus is reached) or fail to resolve it so that, collectively, uncertainty is consequently high. In this case, the issue remained unresolved within the present exercise.

These examples illustrate ways in which important uncertainties in our retrospective re-appraisal of the 1975–76 crises were brought out by adopting a structured elicitation approach.

BBN Parametrization and Evaluation

For the BBN calculations, we transformed the expert group quantile marker values into a standard statistical functional distribution. Two alternative distributions were considered: Dirichlet and Generalized Trapezoidal; and results compared to find which provided the best fit to the elicited joint percentile estimates. Best fits were determined using SolvOpt (Kuntsevich and Kappel 1997), a solver for non-smooth optimization. The solver minimizes a least-squares function: the square of the difference between the solution at each of the three percentiles and the elicited values for the percentiles. Constraints on the distribution parameters are imposed by a penalty function (e.g. setting a large but finite penalty coefficient outside the required bounds of the function). Figure 7 shows the best-fit Dirichlet and Generalized Trapezoidal distributions for nodes 5 (recorded VT rate), 6 (Felt VT rate) and 15 (eruption) alongside the corresponding expert group quantiles. The Dirichlet distribution (i.e. the multivariate Beta) was thus chosen to characterize uncertainty distributions in the final network.

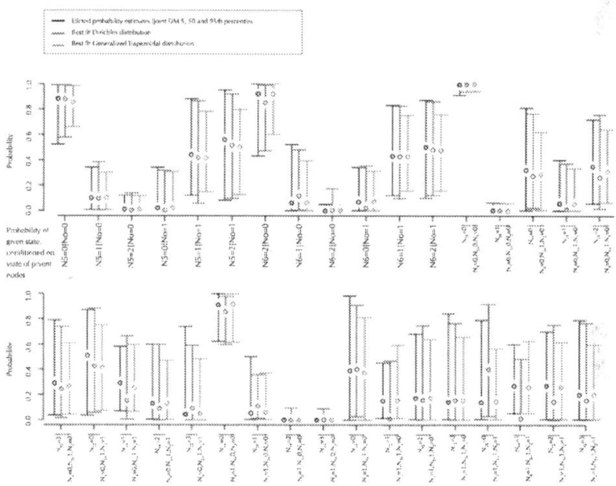

Figure 7: Comparison of the best-fit Dirichlet and Generalized Trapezoidal distributions with expert group quantiles. Comparison of the best-fit Dirichlet and Generalized Trapezoidal distributions for nodes 5 (recorded VT rate), 6 (Felt VT rate) and 15 (eruption), shown alongside the corresponding expert group quantiles. Plot shows conditional probability distributions obtained by fitting Dirichlet distributions (red) and Generalized Trapezoidal distributions (green) to the joint Decision Maker's (DM) 5, 50 and 95th percentile estimates (black; the joint DM being the optimised combination of the individual expert's estimates). The Dirichlet distribution gave the overall best-fit and was chosen for the final network.

In determining the distribution parameters, some logical constraints have been applied. For example, based on past activity the probability of a magmatic blast *P(blast)*at La Soufrière can be expected to be lower than the probability of a magmatic eruption without a blast *P(magmatic eruption)*. However, if the uncertainty spread in the estimate for *P(blast)* is much greater than that of *P(magmatic eruption)*, and the experts gave a broader distribution, this aspect may not be captured properly by fitting a Dirichlet distribution in the standard way with least squares (with all three percentiles weighted equally). To avoid this, we used weights of 1, 1.5, 1 for the 5, 50 and 95%ile respectively – that is, giving a higher weighting to the median value, and imposed the further condition that*P(blast)* < *P*(*magmatic eruption*). This approach more closely reproduced the groups' 'best estimate' values, and creates a logically consistent overall outcome.

Code has been written to perform random sampling from the best-fit Beta (for binary nodes) and Dirichlet distributions for each node using the Scythe Statistical Library (Pemstein et al. 2007). Given the relevant distribution parameters, a set of random samples is returned, corresponding to the probability of each possible outcome (with the total summing to 1). These samples populate the conditional probability tables (CPTs) for each node. As there is no real basis for enforcing correlations between the elicited probability density functions, each distribution was sampled independently. Each run results in a fully-defined CPT for each node, completely parametrizing the network. At this stage we calculate various measures, such as Mutual Information and entropy (described below and in supplementary material [see), which characterize the strength of node relationships and overall uncertainty. The network is updated with the time series of observational evidence (bottom panel in Figure 5), and computes the probability of each unobservable state or outcome (e.g. magmatic eruption), at each time step. The sampling procedure is repeated for a total of 10,000 runs to build output distributions for *P (eruption)*, Mutual Information etc., and the goal being to propagate uncertainty from the experts' initial estimates through to the ultimate query node.

Mutual Information

To investigate the strength of the relationship between any particular node on the BBN and the query node (eruption probability), we compute Mutual Information (MI). The MI of nodes *X* and *Y*, *MI*(*X*, *Y*), is the reduction in uncertainty (entropy) in *Y* by knowing *X*. In a Bayesian Network, higher entropy means the node states are more randomly distributed, therefore more uncertain (see Bedford and Cooke 2001). Zero entropy means the state is known exactly. We can also express the strength of the relationship between nodes as Mutual Information percentage (MI%), the percentage reduction in entropy of node Y given information about X. Zero MI implies conditional independence – node X does not give any information about the state of node Y. MI (and MI%) can be computed for any pair of nodes, regardless of whether they are directly connected. This can be used to assess how individual states and processes might affect hazard outcomes, and as a consequence, impact on risk. Computing this measure for all observables makes it possible to identify objectively which parameters provide the greatest information about future activity. As the BBN presented here has been developed using expert opinion (rather than observational data), MI is a measure of the perceived strength of connection between nodes, and the perceived value of the various observables in forecasting eruptive activity.

Setting the Guadeloupe BBN in Context

The network (Figure 3) was used to infer the various hidden states (e.g. magmatic unrest or ascent) and calculate an evolving probability of eruption using the sequence of observations made between July 1975 and March 1977 (see Figure 5). Before discussing the observation-based probability estimates, we set the scheme in context by looking first at implied recurrence rates for a simple "reference scenario". In this "no signs" scenario (denoted 'no activity' in the table in 5), the assumed situation is that monitoring and observational data are continuously available, but no abnormal activity is detected in any variable. This scenario yields a median probability of magmatic blast or magmatic eruption of the order 10^{-11} for an eruption within a three month interval (i.e. with no unrest evident in the three months prior to eruption), with an upper 95%ile probability of order 10^{-6}. Thus the group assessed the chance of a totally "out-of-the-blue" magmatic eruption (unaccompanied by any precursory signs) within three months to be negligible. The median probability of a phreatic eruption per three month period under the same conditions (i.e. no prior unrest) is much higher, at 0.003 (corresponding to a recurrence rate of approximately one in 80 years), while the expected (mean) probability is 0.01 (1%, a recurrence rate of one in 25 years). Uncertainties are large, and differences between mean and median estimates signal strong skew in the distributions.

During the elicitation exercise it was not feasible to weigh up such end-member recurrence rate uncertainties in detail, not least because there is no real basis on which to calibrate them against geological and historical records. If major discrepancies were suspected, perhaps due to members of the group inflating their rate estimates to reflect their uncertainty in the absence of substantive evidence, this would invite potentially intractable debate. Faute de mieux, the rate estimates are reproduced here to illustrate how even extremely unlikely, exceptional scenarios can be accommodated in a BBN, albeit with gross uncertainties; the numbers are not definitive.

RESULTS AND DISCUSSION

Inferring Eruption Probabilities using 1975–1977 Observations

Figure 5 shows the sequence of estimated probabilities over the following three months for: (a) a magmatic eruption or blast, (b) phreatic eruption, or (c) no eruption, with associated uncertainties in these estimates. The solid black lines show the expected (mean) probability estimates, the broken lines indicate the medians (50th percentile), with the shaded areas outlining the 5–95 percentile ranges.

The data are presented in From July 1975 until early July 1976, the likelihood of a magmatic eruption was only very marginally elevated above the assumed base-rate: at the 95%ile level this probability would have been less than 0.01 (i.e. only a 5% chance it would be greater), while the corresponding 95%ile probability for a phreatic explosion was 0.07. This very limited increase in probability at the start of July 1976 indicates the influence of false positive and true negative rates for signs associated with magmatic eruption, including VT activity. If post-hoc this appears surprisingly low it serves to illustrate the challenge of setting up a CPT with appropriate diagnostic powers for the various precursors when these are poorly known *a priori*, and can also reflect the fact that some observations can run counter to others. Perhaps even more pertinently, it illustrates potential pitfalls with hindsight interpretation of evidence: some scientists present in 1976 were quite convinced a magmatic eruption was not going to happen.

From 9 July 1976 onwards, the perceived median probability of a magmatic eruption or magmatic blast within three months increased to 0.3, the probability of a phreatic eruption also increased (to around 0.1) [see 5] and the most likely outcome was still "no eruption". The occurrence of the explosion on 8 July 1976 did not modulate future event probabilities as these were solely conditional on observational data trends. In other words, the likelihood of another explosion in the following three months did not change in the wake of one that had just happened - incorporating this additional dependence would improve the BBN model.

The probability of a magmatic eruption fell back marginally at 13 August 1976, but rose again to peak at the end of the month when fresh glass was reported as having been detected in a microscope sample and all volcanic other phenomena were observed as present or elevated (see lower panels, Figure 5). Magmatic eruption probability remained elevated until the end of November 1976. Thereafter the states of a number of observables dropped (see Figure 5, lower panels) and the probability of eruption fell noticeably. However, it still remained above the pre-July level, and stayed there through until March 1977 (at which time accessible data became incomplete and our analysis was curtailed). From early July 1976, the likelihood of a phreatic explosion remained elevated at a roughly constant level, reducing slightly through March 1977. Whilst uncertainty in magmatic and phreatic eruption probability was higher after 29 August 1976 than before 9 July [see 5] , lower bound estimates returned close to pre-July levels, with a corresponding shift up in the upper bound likelihood of no eruption.

The corresponding marginal distributions for probability of magmatic intrusion and ascent are given in 6. The median probability of magmatic unrest with no precursory activity is estimated to be of the order 10^{-8} (mean 3×10^{-4}), which can be compared to the 'baseline' elicited probability of magmatic unrest (the CPT for node 0) - a median value of 0.08, or 1 episode every 3.1 years. In considering this estimate, one hypothesis is that exceptionally elevated VT swarm activity is evidence of magma intrusion rather than abnormal non-magmatic hydrothermal perturbation. If we deem ~15 VTs/ month to represent the baseline seismic activity related to a steadily degassing magma reservoir, observing a VT rate well above this might be taken as evidence of a switch due to upward magma intrusion from the shallow chamber. Applying this logic to the period before 1975, we find six potential instances of possible intrusion between 1956 and 1975 (including that at the beginning of the 1975–77 crises, in July 1975). Six intrusions in 20 years imply one every 3.3 years, largely in line with the elicited joint median value of 1 event per 3.1 years. There is, however, large uncertainty in the experts' joint probability estimate – the corresponding 5 and 95 percentile estimates are 7.6×10^{-5} (1 event per ~3300 years) and 0.81 (just over 3 episodes / year).

This hypothesis presumes that all phreatic eruptions are still-born failed magmatic eruptions, and that magma intruded towards the surface heated the hydrothermal system, pressurized it, opened fractures and generated VTs in competent rock, but then stalled at shallow depth. More recent occurrences of such "intrusive swarms", if that is what they are, suggest almost continuous intrusive activity from 1992, ramping up to 1997–98 when sudden and sustained chlorine degassing appeared and peaked. This is compatible with a phase of magma intrusion into the dome system, with declining degassing at its periphery but much increased gas and heat flux at the summit. An earlier, notable pulse of "intrusive swarms" occurred in 1962–1968; magma intrusion at that time is inferred by Ruzié et al. (2012), based on Noble gas isotope systematics and other data.

The question of whether all pre-1976 historical phreatic eruptions were magmatic in origin is moot. An example of an alternative hypothesis (not considered in our study) is that gas fluxes in magma reservoirs may be decoupled and independent of magma fluxes. Thus episodes of unrest record the transfer of gas to higher levels unaccompanied by magma. The great value of a structured probabilistic elicitation approach is that it helps raise scientific questions of this kind, and encourages development of causal conceptual models that can be tested with observations and theoretical models. The preferred conceptual model can then be used to structure a BBN, dynamically trained on the available observations (and other information) and then used to evaluate scenario probabilities for new unrest situations.

BBN Mutual Information

The BBN formulation allows analytic testing of node state conditionalities and uncertainty in the network. For instance, entropy is a measure of unpredictability in a system (Bedford and Cooke2001), with higher entropy indicating the node states are more randomly distributed, and therefore more uncertain. Here we restrict our investigation to the strength of the relationship between individual nodes in the BBN and the query node, and compute corresponding Mutual Information percentages MI% for each pair (Ebert-Uphoff 2007; see Methods and 4). MI% is a measure of the reduction of uncertainty in the target node (eruption) due to knowledge of another node; either an observational node or an inferred latent (hidden) node. This allows us to identify which monitoring parameters provide the greatest information about future activity.

Figure 8 summarizes MI% results: black diamond's denote the mean MI % estimate, with boxes denoting the median (centreline) and first and third quantiles; whiskers on the plot depict 5th and 95th percentiles. In terms of eruption probability, the four most important nodes (0, 2, 3 and 4 – see above for definitions) are all latent factors concerning aspects of magma condition or behaviour at depth. These cannot be observed directly and can only be inferred from related observations. Deep source deformation (node 1) is a latent node with moderate potential to influence eruption probability.

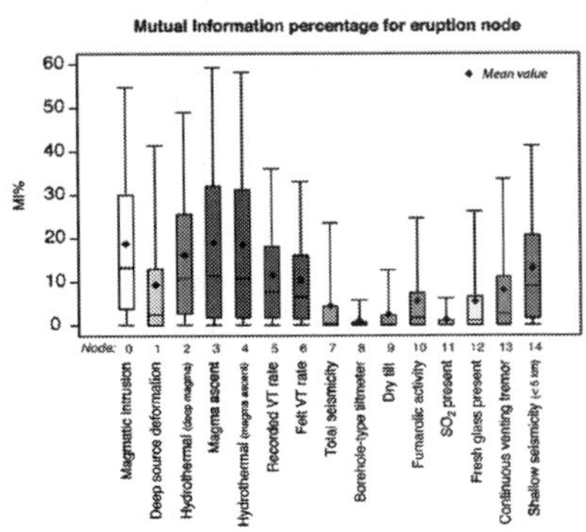

Figure 8: Mutual Information. Mutual Information (MI) is a measure of the strength of the relationship between two nodes. By computing the MI for

eruptive outcome with each observable, it is possible to identify which parameters provide the greatest information about future activity. The black diamond's denote the mean MI % estimate. Boxes denote the median (centreline), first and third quantiles. The whiskers on the plot denote the 5th and 95th percentiles. The data are provided in .

In 1976, the most informative observational nodes in relation to eruption potential are inferred to be seismicity (i.e. nodes 5, 6, 13 and 14), with shallow seismicity above 5 km depth identified as having the strongest indicative power. In the BBN, shallow seismicity is closely associated with inferences about magma ascent, and therefore acts to influence the query node through its links to those latent nodes (e.g. nodes 3 and 4). The 1976 tilt measurements (nodes 8 and 9) and measurements of SO_2 (node 11) were of little significance for assessing eruption likelihood. These findings might be less applicable in current times with modern techniques, better instruments, and advances in data interpretation and physical models.

Identification of fresh glass (node 12) can be a valuable petrological indicator, but turned out to be a false positive here. Properly, the presence or absence of fresh glass is not a single binary yes/no parameter and its evidential worth should be considered in conjunction with other observables. As Cashman and Hoblitt (2004) have shown, petrological evidence from early erupted products can be highly diagnostic. However, in an escalating eruption, obtaining samples can be challenging and risky and dedicated resources are needed to perform fast petrological analysis. Rapid interpretation with a simple binocular microscope requires experience, care and awareness of context. Here the BBN formulation is indispensable – the implications should be modulated by the reliability of the observation. In the escalating sequence at Merapi 2010, for instance, evidence for vesicular material in the very earliest tephra was found in material erupted three days before the dome was first seen and then again in material before the paroxysmal dome explosion on 5 November, however the analysis was not performed until some months after the events unfolded (Komorowski et al. 2013). The example of fresh glass highlights the vital importance of characterizing fully both the diagnostic sensitivity and specificity (e.g. Sackett et al. 1996) when weighing strands of scientific evidence.

The wide uncertainty spreads shown on Figure 8 (see also 7) illustrate clearly just how weak, individually, these various indicators were for eruption forecasting on Guadeloupe in 1976. Calculation of Mutual Information allows parameters that have poor diagnostic value to be identified, potentially enabling resources to be focussed on more diagnostic observations and allowing simplification of the BBN by removing links which emerge as uninformative.

DISCUSSION AND CONCLUSIONS

The findings of this exploratory Bayesian Belief Network analysis of the 1976 Guadeloupe crisis lend objective support to the retrospective view that the authorities' concern for public safety and decision to evacuate were rational and defensible, given all the major scientific uncertainties that existed at the time. It is clear that even now, despite 40 years of intervening advances in research, the evidence of the time engenders considerable concern as well as uncertainty. Highly definite views, expressed at the time by some scientists, are hard to justify given the large uncertainties in the evidence, and the official precautionary decision to evacuate, given the potential consequences of an explosive magmatic eruption, is not weakened by this hindcasting analysis. The BBN analysis also helps highlight some of the challenges of providing science-based decision support under conditions of considerable uncertainty that will emerge again in a future volcanic crisis, on Guadeloupe or elsewhere.

In the present case, some of our retrospective probability estimates may be unduly or conservatively high (for example, the relatively high probability of magmatic blast as compared to other, lower intensity outcomes). In part, this may be due to the restricted scope and rather perfunctory nature of the elicitation; a more comprehensive exercise, with more experts, would be desirable. However, the BBN calculations illustrate the substantial uncertainties that are typically associated with interpreting incomplete observational information or monitoring data of limited quality. Volcanology has moved on since 1976, of course, and many monitoring techniques have improved immensely. This said, the basic evidential principles outlined here remain the same, and demonstrate how crucial it is, when resources are limited, to focus monitoring efforts on those parameters which maximise strength of inference about key hidden conditions and latent factors, such as magma ascent.

The information that goes into building the BBN model can offer considerable help in expressing and communicating scientific uncertainties and their sources, and in elucidating how forecast outcomes are sensitive to different assumptions and relationships. For example, the mutual information measure of conditional dependency between two variables is an elegant way of identifying and quantifying dependencies between elements in a system that may not be immediately obvious, especially in a complex situation involving a model with numerous inputs, nodes, and interactions.

Assessing and presenting uncertainty to policy- and decision-makers has emerged as an important topic in all branches of hazard and risk analysis, not least for the situation where we are confronted by escalating volcanic unrest (for a review of approaches to volcanic hazard assessment, see Marzocchi and Bebbington 2012). Due to the intrinsic process complexities and limitations in our understanding (present in all natural hazards domains), volcano forecasting assessments are inevitably imprecise and a sense of vagueness can be communicated through the use of confidence ranges.

Recent work by Dieckmann et al. (2010) suggests that decision makers are not necessarily 'ambiguity averse' in a forecast context and presenting ranges of probability can have distinct advantages as a way to communicate probability and diagnostic confidence. The study by Dieckmann et al. (in a military intelligence context) indicated a smaller ambiguity effect in decisions that were taken when a narrative forecast was accompanied by a probability range as opposed to the same narrative with simply a point value. However, in one of their tests, it was thought that the point estimate was more useful for decision making at low probabilities. Dieckmann et al. also found that when evaluating a forecast in hindsight, their decision makers tended to report lower levels of blame and higher levels of source credibility for forecasts that reported uncertainty ranges as compared to single value point assessments. How these findings might map across to volcanic hazard forecasts and decision support situations requires further research.

Future Developments

This retrospective analysis is necessarily simple in scope, reflecting the limited information available during the 1976 crisis. This said, the basic framework can easily be developed to reflect current scientific understanding and to incorporate a much wider range of monitoring data and observations. The diversification of new volcano monitoring techniques is such that a formulation for weighing and pooling multiple strands of observational evidence is becoming indispensable, especially if an audit trail is required to track science-based inputs to decisions under rapidly-changing conditions. Such a network can be utilised as an automated tool for real-time use in the volcano observatory, using streams of monitoring data to generate and continuously update probabilistic hazard forecasts.

The network presented here is a static BBN that weighs the evidence in discrete and unconnected time steps. With sufficient knowledge of the volcanic system and with comprehensive, repeated observational data, more sophisticated Dynamic Bayesian Networks (DBNs) can be constructed to model temporal relationships between nodes (Pearl 2000; Murphy 2002; Jensen and Graven-Nielsen2007).

Whereas a static BBN describes the state of a system without using information about its prior history, a dynamic BBN can incorporate crucial information about system evolution in which the state of the volcano at any time is dependent on any number of past states. The order of such a model is the length of history, or 'memory' of the processes concerned. Dynamic BBN nodes can be tied over many time-slices to represent higher order processes, as appropriate.

One volcanological example of a DBN is the Hidden Multi-state Markov Model (HMM) of Aspinall et al. (2006). The HMM is a simple case of a DBN with a single discrete hidden node (e.g. see Rabiner1989; Jensen 1996). The HMM model is based on the assumption that multi-parameter monitoring data can be jointly evaluated to infer time to eruption, and hence used to inform hazard alert levels. For data-rich applications, network parameters and even the network structure itself can be estimated from data, using learning algorithms (Murphy 2002). Such an extension is presented in Hincks et al. (2006), which apply learning algorithms to parameterize a DBN for forecasting dome collapse on Montserrat using multiple time series of monitoring data. Determining network structure from time-evolving data is an advanced technique which would be especially salient for observational volcanology in an unrest crisis; this theme will be developed elsewhere.

AUTHORS' CONTRIBUTIONS

The BBN structure was developed through discussions between the authors and all four contributed to writing the paper. R.S.J.S., W.P.A. and J-C.K. advised on the key volcanic processes and participated in the expert judgement elicitation. T.K.H. wrote the code and performed the the BBN analysis and network parameter fitting, and drafted the diagrams (with the exception of Figures 5 and 6, produced by W.P.A.). W.P.A. ran EXCALIBUR for processing elicitation responses, analysed the results of the expert elicitation; he also provided direct knowledge about monitoring and events during the 1976 Guadeloupe crisis. J-C.K. provided background information on the eruptive history, the chronology of the crisis (Tables 1 and 2), and monitoring data for La Soufrière. All authors read and approved the final manuscript.

ACKNOWLEDGEMENTS

The authors would like to thank A. Rust, J. Gottsmann and Y. Legendre for participating in the expert elicitation and their contributions to discussions. This work was supported by VOLDIES (Advanced European Research Council grant to RSJ Sparks). EU Project EXPLORIS (EVR1-2001-00047) and Aspinall & Associates partly funded early stages of this work. Support was provided by the Institut National des Sciences de l'Univers (INSU-CNRS) and the CASAVA Project (ANR-09-RISK-02, J.C. Komorowski). JCK is grateful for valuable and insightful discussions with F. Beauducel. We gratefully acknowledge the original intellectual stimulation and advice of Dr Gordon Woo, who encouraged us to undertake this study.

REFERENCES

1. Annen C, Wagner J-J (2003) the impact of volcanic eruptions during the 1990s. Nat Hazards Rev 4(4):169-175
2. Aspinall WP (2006) Structured elicitation of expert judgement for probabilistic hazard and risk assessment in volcanic eruptions. In: Mader HM, Coles SG, Connor CB, Connor LJ (eds) Statistics

in Volcanology. Special Publications of IAVCEI, 1, London: Geological Society. pp 15-30

3. Aspinall WP, Cooke R (2013) Expert Elicitation and Judgement. In: Hill L, Rougier JC, Sparks RSJ (eds) Risk and Uncertainty assessment in natural Hazards, Cambridge: Cambridge University Press.
4. Aspinall WP, Woo G (1994) an impartial decision-making procedure using expert judgement to assess volcanic hazards. Atti dei Convegni Lincei 112:211-220
5. Aspinall WP, Sigurdsson H, Shepherd JB (1973) Eruption of Soufrière volcano on St. Vincent Island, 1971–1972. Science 181:117-124
6. Aspinall WP, Michael MO, Tomblin JF (1976) Evidence for fluid bodies beneath the Sulphur Springs Geothermal Region, St. Lucia, West Indies. Geophys Res Lett 3:87-90
7. Aspinall WP, Loughlin SC, Michael FV, Miller AD, Norton GE, Rowley KC, Sparks RSJ, Young SR (2002) The Montserrat Volcano Observatory: its evolution, organisation, role and activities. In: Druitt TH, Kokelaar BP (eds) The eruption of Soufrière Hills Volcano, Montserrat, from 1995 to 1999, London: Geological Society. pp 71-92 Memoir No. 21
8. Aspinall WP, Woo G, Voight B, Baxter P (2003) Evidence-based volcanology: application to eruption crises. J Volcanol Geotherm Res 128(1–3):273-285
9. Aspinall WP, Carniel R, Jaquet O, Woo G, Hincks T (2006) Using hidden multi-state Markov models with multi-parameter volcanic data to provide empirical evidence for alert level decision-support. J Volcanol Geotherm Res 153:112-124
10. Barberi F, Bertagnini A, Landi P, Principe C (1992) A review on phreatic eruptions and their precursors. J Volcanol Geotherm Res 52:231-246 doi:10.1016/0377-0273(92)90046-G
11. Beauducel F (2006) À propos de la polémique de Soufrière 1976. http://www.ipgp.fr/~beaudu/soufriere/forum76.html *webcite*. Accessed 21 December 2013
12. Bedford T, Cooke R (2001) Probabilistic Risk Analysis. Foundations and Methods: Cambridge University Press.

13. Blérald A-P (1986) Histoire éruptive de la Guadeloupe et de la Martinique du XVIIème siècle à nos jours. Paris: Editions Karthala.
14. Boichu M, Villemant B, Boudon G (2011) Degassing at La Soufriere de Guadeloupe volcano (Lesser Antilles) since the last eruptive crisis in 1975–77: Result of a shallow magma intrusion? J Volcanol Geotherm Res 203(3–4):102-112
15. Borsuk ME, Stow CA, Reckhow K (2003) Integrated approach to total maximum daily load development for Neuse River Estuary using Bayesian probability network model (Neu-BERN). J Water Res Pl-ASCE 129(4):271-282
16. Boudon G, Semet P, Vincent PM (1987) Magma and hydrothermally driven sector collapses: the 3100 and 11500 y B.P. eruptions of La Grande Découverte (La Soufrière) Volcano, Guadeloupe, and French West Indies. J Volcanol Geotherm Res 33:317-323
17. Boudon G, Dagain J, Semet M, Westercamp D (1988) Carte géologique à 1/20000e du Massif volcanique de la Soufrière, 1 sheet, Notice explicative de la carte géologique 1/20000e du Massif volcanique de la Soufrière - Carte Géologique. Editions BRGM, Orléans: BRGM-CNRS-DRM-IPGP. pp 1-43
18. Boudon G, Le Friant A, Komorowski J-C, Deplus C, Semet MP (2007) Volcano flank instability in the Lesser Antilles Arc: diversity of scale, processes, and temporal recurrence. J Geophys Res B 112:B08205-B08232
19. Boudon G, Komorowski J-C, Villemant B, Semet MP (2008) A new scenario for the last magmatic eruption of La Soufriere of Guadeloupe (Lesser Antilles) in 1530 AD Evidence from stratigraphy radiocarbon dating and magmatic evolution of erupted products. J Volcanol Geotherm Res 178(3):474-490
20. Brousse R, Monneyron N, Semet MP (1977) Sur la présence de verres non altérés dans les projections de la Soufrière (Guadeloupe), durant la crise de 1976. CR Acad Sci, Paris, Sér D 285:753-754
21. Carlut J, Quidelleur X, Courtillot V, Boudon G (2000) Paleomagnetic directions and K/Ar dating of 0 to 1 Ma lava flows from La Guadeloupe Island (French West Indias): Implications for time-averaged field models. J Geophys Res B 105(B1):835-849

22. Cashman KV, Hoblitt RP (2004) Magmatic precursors to the 18 May 1980 eruption of Mount St. Helens, USA. Geology 32:141-144
23. Cooke RM (1991) Experts in Uncertainty - Opinion and Subjective Probability in Science. Environmental Ethics and Science Policy. Series: Oxford University Press.
24. Cooke R, Kritchallo V, Solomatine D (2000) EXCALIBUR v1.0 Delft University of Technology, the Netherlands.http://risk2.ewi.tudelft.nl/oursoftware/6-excalibur *webcite*. Accessed 17 July 2013
25. Cooke RM, Kurowicka D, Hanea AM, Morales O, Ababei DA, Ale B, Roelen A (2007) Continuous/Discrete Non Parametric Bayesian Belief Nets with UNICORN and UNINET, Proceedings of Mathematical Methods in Reliability MMR 2007. UK: Glasgow.
26. Cowell R, Verrall R, Yoon Y (2007) Modeling Operational Risk with Bayesian Networks. J Risk Insur 74(4):795-827 doi:10.1111/j.1539-6975.2007.00235.x
27. De Vanssay B (1979) Les événements de 1976 en Guadeloupe: apparition d›une subculture du désastre. PhD thesis, Centre Universitaire Antilles-Guyane (Pointe-à-Pitre, Guadeloupe) ET Ecole des Hautes Etudes en Sciences Sociales, Université Paris 5.
28. Decision Systems Laboratory (2013) GeNIe (Graphical Network Interface) software. University of Pittsburgh. http://genie.sis.pitt.edu *webcite* Accessed 17 Dec 2013
29. Dieckmann NF, Mauro R, Slovic P (2010) The effects of presenting imprecise probabilities in intelligence forecasts. Risk Anal 30:987-1001 doi:10.1111/j.1539-6924.2010.01384.x
30. Dorel J, Feuillard M (1980) Note sur la crise sismo-volcanique à la Soufriére de la Guadeloupe 1975–1977. Bull Volcanol 43(2):419-430
31. Ebert-Uphoff I (2007) Users Guide for the LinkConnectionStrength Package (Version 1.0) – A PNL Package for the Calculation and Visualization of Entropy, Link Strengths and Connection Strengths in Discrete Bayesian Networks.http://www.dataonstage.com/PNL/PACKAGES/PTG_LCS/PTG_LCS.html *webcite*. Accessed 5 June 2013

32. Feuillard M (2011) La Soufrière de Guadeloupe: UN volcan et un peuple. 1-246
33. Feuillard M, Allegre C, Brandeis G, Gaulon R, Le Mouel JL, Mercier J, Pozzi J, Semet MP (1983) The 1975–1977 crisis of La Soufrière de Guadeloupe (F.W.I): A still-born magmatic eruption. J Volcanol Geotherm Res 16:317-334
34. Fiske R (1984) Volcanologists, journalists, and the concerned local public: a tale of two crises in the eastern Caribbean. In: Boyd F (ed) Explosive Volcanism: Inception, Evolution and hazards. Studies in Geophysics, Washington DC: National Academy Press. pp 170-176
35. Heiken G, Crowe B, McGetchin T, West F, Eichelberger J, Bartram D, Peterson R, Wohletz K (1980) Phreatic eruption clouds: the activity of la Soufrière de Guadeloupe, F.W.I., August — October, 1976. Bull Volcanol 43(2):383-395 doi:10.1007/BF02598039
36. Hincks T (2007) Probabilistic volcanic hazard and risk assessment. PhD thesis: University of Bristol.
37. Hincks T, Aspinall W, Woo G (2006) Evidence-based real-time hazard evaluation methodology and software for EXPLORIS volcanoes. Deliverable D6.4 for the project Explosive Eruption Risk and Decision Support for EU Populations Threatened by Volcanoes (EXPLORIS) EU Project EVR1-2001-00047.Available from http://exploris.pi.ingv.it/confidential/products/index.html *webcite*. Accessed 17 July 2013
38. Hirn A, Michel B (1979) Evidence of migration of main shocks during major sismovolcanic crisis of la Soufriére (Guadeloupe, Lesser Antilles) in 1976. J Volcanol Geotherm Res 6:295-304
39. IPGP (1956–2013) Rapport d'activité de l'Observatoire Volcanologique et Sismologique de Guadeloupe, Catalogue de la sismicité instrumentale enregistrée en Guadeloupe, et Bilan Mensuel de l'activité Volcanique de la Soufrière et de la Sismicité régionale. Observatoire Volcanologique et Sismologique de la Guadeloupe, Institut de Physique du Globe de Paris, IPGP-CNRS-INSU, Conseil, Général de Guadeloupe. published online at: http://www.ipgp.fr/pages/0303040901.php *webcite* accessed 17 July 2013; unpublished database, available upon request at http://www.ipgp.fr *webcite* andhttp://centrededonnees.ipgp.fr/adresse.php *webcite*

40. Jensen FV (1996) An Introduction to Bayesian Networks. London: Taylor and Francis.
41. Jensen F, Graven-Nielsen T (2007) Bayesian Networks and Decision Graphs (Information Science and Statistics). New York: Springer.
42. Jolivet J (1958) La Crise Volcanique de 1956 à La Soufrière de La Guadeloupe. Ann Géophys 14(3):305-322
43. Kokelaar BP (2002) Setting, chronology and consequences of the eruption of Soufrière Hills Volcano, Montserrat (1995–1999). In: Druitt TH, Kokelaar BP (eds) The eruption of Soufrière Hills Volcano, Montserrat, from 1995 to 1999, 1st edn. pp. 1-44
44. Komorowski J-C, Boudon G, Semet M, Samper A, Quidelleur X, Coudret E, Le Friant A, Venat J, Paterne M, Villemant B, Beauducel F, Hammouya G (2004) Report on the eruptive history of Soufrière of Guadeloupe volcano in the last 100,000 years based on field, analytical, and bibliographic studies. Deliverable D2.5 for the project Explosive Eruption Risk and Decision Support for EU Populations Threatened by Volcanoes (EXPLORIS) EU Project EVR1-2001-00047. http://exploris.pi.ingv.it/confidential/products/Report/DELIVERABLES/deliverable_annuale 2003. htm*webcite* Accessed 17 July 2013
45. Komorowski J-C, Boudon G, Semet MP, Beauducel F, Anténor-Habazac C, Bazin S, Hammouya G (2005) Guadeloupe. In: Lindsay JM, Robertson REA, Shepherd JB, Ali S (eds) Volcanic Hazard Atlas of the Lesser Antilles. Seismic Research Unit of the University of the West Indies, pp. 68-107
46. Komorowski J-C, Legendre Y, Caron B, Boudon G (2008) Reconstruction and analysis of sub-Plinian tephra dispersal during the 1530 A.D. Soufrière (Guadeloupe) eruption: implications for scenario definition and hazards assessment. J Volcanol Geotherm Res 178:491-515
47. Komorowski J-C, Boudon G, Le Friant A, and Legendre Y (2008b) A remarkable Holocene record of flank-collapses at La Soufrière volcano (Guadeloupe): implications for future hazards and scenarios. In: Paper presented at the International Association of Volcanology and Chemistry of the Earth's Interior (IAVCEI). Reijkavik, Iceland: General Assembly. pp 17-23 2008

48. Komorowski J-C, Jenkins S, Baxter P, Picquout A, Lavigne F, Charbonnier SJ, Gertisser R, Preece K, Cholik N, Budi-Santoso A, Surono (2013) Stratigraphy and facies relationships of explosion-generated pyroclastic density current deposits from the Merapi 2010 eruption. J Volcanol Geotherm Res 261:260-294
49. Kuntsevich A, Kappel F (1997) Solvopt (Solver for local optimization problems). Software.http://www.uni-graz.at/imawww/kuntsevich/solvopt/ *webcite*. Accessed 17 July 2013
50. Le Guern F, Bernard A, Chevrier RM (1980) Soufrière of Guadeloupe, 1976–1977 eruption mass and energy transfer and volcanic health hazards. Bull Volcanol 43:577-594
51. Legendre Y (2012) Reconstruction fine de l'histoire éruptive et scénarii éruptifs à la Soufrière de Guadeloupe: vers un modèle intégré de fonctionnement du volcan. PhD thesis. Paris: IPGP.
52. LegendreY, Komorowski J-C, Boudon G (2010) A multidisciplinary approach for high-resolution reconstruction of the eruptive past of La Soufrière (Guadeloupe) over the last 12 000 years: Implications for hazard assessment 12: Vienna: European Geophysical Union General Assembly 2010, pp EGU2010-EGU12719 02–07 May 2010
53. Lepointe E (1999) Le réveil du volcan de la Soufrière en 1976: la population guadeloupéenne à l'épreuve du danger. In: Yacou A (ed) Les catastrophes naturelles aux Antilles – D'une Soufrière à l'autre, Karthala, Paris: CERC Université Antilles et de la Guyane. pp 15-71
54. Li L, Jendrzejewski N, Aubaud C, Bonifacie M, Crispi O, Dessert C, Agrinier P (2012) Isotopic evidence for quick freshening of magmatic chlorine in the Lesser Antilles arc volcanoes. San Francisco, USA: Abstract presented at AGU Fall Meeting, 3–5 December 2012.980 V53B-2831
55. Loubat B, Pistolesi-Lafont A (1977) La Soufrière - à qui la faute?. Paris: Presses de la Cité.
56. Marinelli G (1976) Sur la «cendre» projetée per la Soufrière (Guadeloupe) en août 1976. CR Acad Sci Paris, Sér D 283:1731-1732
57. Marzocchi W, Bebbington MS (2012) Probabilistic eruption forecasting at short and long time scales. Bull Volcanol 74(8):1777-1805 doi:10.1007/s00445-012-0633-x

58. Marzocchi W, Sandri L, Gasparini P, Newhall C, Boschi E (2004) Quantifying probabilities of volcanic events: The example of volcanic hazard at Mount Vesuvius. J Geophys Res B 109:B11201. doi:10.1029/2004JB003155
59. Marzocchi W, Sandri L, Furlan C (2006) A quantitative model for volcanic hazard assessment. In: Mader HM, Coles SG, Connor CB, Connor LJ (eds) Statistics in Volcanology. Special Publications of IAVCEI, 1, London: Geological Society. pp 31-37
60. Marzocchi W, Sandri L, Selva J (2008) BET_EF: a probabilistic tool for long- and short-term eruption forecasting. Bull Volcanol 70(5):623-632
61. Molina M, Fuentetaja R, Garrote L (2005) Hydrologic Models for Emergency Decision Support Using Bayesian Networks. Lecture Notes in Computer Science: Symbolic and Quantitative Approaches to Reasoning with Uncertainty 3571:88-99 doi:10.1007/11518655_9
62. Murphy KP (2002) Dynamic Bayesian Networks: Representation, Inference and Learning. Ph.D. thesis. UC Berkeley: Computer Science Division.
63. Nakada S, Shimizu H, Ohta K (1999) Overview of the 1990–1995 eruption at Unzen Volcano. J Volcanol Geotherm Res 89:1-22 doi:10.1016/S0377-0273(98)00118-8
64. Neil M, Fenton N, Tailor M (2005) Using Bayesian Networks to Model Expected and Unexpected Operational Losses. Risk Anal 25(4):963-972 doi:10.1111/j.1539-6924.2005.00641.x
65. Newhall C, Hoblitt R (2002) Constructing event trees for volcanic crises. Bull Volcanol 64(1):3-20 doi:10.1007/s004450100173
66. Niculescu RS, Mitchell TM, Rao RB (2006) Bayesian network learning with parameter constraints. J Mach Learn Res 7:1357-1383
67. Norsys (2013) Netica (Bayesian Network development software). http://www.norsys.com/ *webcite* Accessed 17 Dec 2013
68. Pearl J (2000) Causality: Models, Reasoning, and Inference. Cambridge: Cambridge University Press.
69. Pemstein D, Quinn KM, Martin AD (2007) Scythe Statistical Library: Application Programmers' Interface.http://scythe.wustl.edu/api/ *webcite*. Accessed 17 July 2013

70. Préfecture de Guadeloupe (1977) Volcan de la Soufrière en Guadeloupe: les événements de 1976. Janvier: Service d›information de la Préfecture de Guadeloupe.1977.
71. Rabiner L (1989) A tutorial on Hidden Markov-Models and selected applications in speech recognition. P IEEE 77(2):257-286 doi:10.1109/5.18626.
72. Ruzié L, Moreira M, Crispi O (2012) Noble gas isotopes in hydrothermal volcanic fluids of La Soufrière of Guadeloupe (Lesser Antilles). Chem Geol 304–305:158-165
73. Sackett D, Rosenberg W, Gray J, Haynes R, Richardson W (1996) Evidence based medicine: What it is and what it isn't - It's about integrating individual clinical expertise and the best external evidence. Brit Med J 312(7023):71
74. Samper A, Quidelleur X, Lahitte P, Mollex D (2007) Timing of effusive volcanism and collapse events within an oceanic arc island: Basse-Terre, Guadeloupe archipelago (Lesser Antilles Arc). Earth Planet Sc Lett 258(1–2):175-191 doi:10.1016/j.epsl.2007.03.030
75. Samper A, Quidelleur X, Komorowski J-C, Lahitte P, Boudon G (2009) Effusive history of the Grande Découverte Volcanic Complex, southern Basse-Terre (Guadeloupe, French West Indies) from new K-Ar Cassignol-Gillot ages. J Volcanol Geotherm Res 187(1–2):117-130
76. Shepherd JB, Aspinall WP, Rowley KC, Pereira J, Sigurdsson H, Fiske RS, Tomblin JF (1979) The eruption of Soufrière volcano, St Vincent April–June 1979. Nature 282:24-28
77. Sheridan MF (1980) Pyroclastic block flow from the September, 1976, eruption of La Soufrière volcano. Guadeloupe, Bull Volcanol 43:397-402
78. Siebert L, Simkin T (2002–2011) Volcanoes of the World: an Illustrated Catalog of Holocene Volcanoes and their Eruptions. Smithsonian Institution, Global Volcanism Program Digital Information Series, GVP-3.http://www.volcano.si.edu *webcite* Accessed 17 July 2013
79. Smithsonian Institution (1976) Soufrière Guadeloupe. Scientific Event Alert Network (SEAN) Bulletin 01:12, September 1976.

80. Smithsonian Institution (1977) Soufrière Guadeloupe. Scientific Event Alert Network (SEAN) Bulletin 02:02, February 1977.

81. Sobradelo R, Martí J (2010) Bayesian event tree for long-term volcanic hazard assessment: Application to Teide - Pico Viejo stratovolcanoes, Tenerife, Canary Islands. J Geophys Res 115:B05206 doi: 10.1029/2009JB006566.

82. Sparks RSJ, Aspinall WP (2004) Volcanic Activity: Frontiers and Challenges in Forecasting, Prediction and Risk Assessment. In: Sparks RSJ, Hawkesworth CJ (eds) The State of the Planet: Frontiers and Challenges in Geophysics, Washington, D. C.: American Geophysical Union.doi:10.1029/150GM28.

83. Spiegelhalter DJ, Dawid A, Lauritzen SL, Cowell R (1993) Bayesian-Analysis in Expert-Systems. Stat Sci 8(3):219-247

84. Stassopoulou A, Petrou M, Kittler J (1998) Application of a Bayesian network in a GIS based decision making system. Int J Geogr Inf Sci 12(1):23-46 doi: 10.1080/136588198241996.

85. Touboul M, Bourdon B, Villemant B, Boudon G, Joron J-L (2007) 238U-230Th- 226Ra disequilibria in andesitic lavas of the last magmatic eruption of Guadeloupe Soufrière, french Antilles: Processes and timescales of magma differentiation. Chem Geol 246(3–4):181-206

86. Villemant B, Hammouya G, Michel A, Semet MP, Komorowski J-C, Boudon G, Cheminee JL (2005) The memory of volcanic waters: Shallow magma degassing revealed by halogen monitoring in thermal springs of La Soufrière volcano (Guadeloupe, Lesser Antilles). Earth and Planet Sc Lett 237(3–4):710-728

87. West FG, Heiken GH, Homuth EF, Peterson RW, Crowe BM, Wohletz KH (1976) Tilts associated with Volcanic Activity, Guadeloupe, French West Indies. Informal report LA-7500-MS. NM, USA: Tech rep Los Alamos Scientific Laboratory.

88. Woo G (2008) Probabilistic criteria for volcano evacuation decisions. Nat Hazards 45(1):87-97 doi: 10.1007/s11069-007-9171-9.

89. Young SR, Sparks RSJ, Aspinall WP, Lynch LL, Miller AD, Robertson REA, Shepherd JB (1998) Overview of the eruption of Soufriere Hills Volcano, Montserrat, 18 July 1995 to December 1997. Geophys Res Lett 25:3389-3392

90. Zlotnicki J, Boudon G, Le Moüel J-L (1992) the volcanic activity of La Soufrière of Guadeloupe (Lesser Antilles): structural and tectonic implications. J Volcanol Geotherm Res 49:91-104

91. Zlotnicki J, Vargemezis G, Mille A, Bruère F, Hammouya G (2006) State of the hydrothermal activity of Soufrière of Guadeloupe volcano inferred by VLF surveys J Appl Geophys 58(4):265-279

Chapter 6

Physical Property Relationships of the Rotokawa Andesite, a Significant Geothermal Reservoir Rock in the Taupo Volcanic Zone, New Zealand

Paul A Siratovich[1], Michael J Heap[2], Marlène C Villenueve[1], James W Cole[1], and Thierry Reuschlé[2]

[1]Department of Geological Sciences, University of Canterbury, Private Bag 4800, Christchurch 8140, New Zealand

[2]Laboratoire de Déformation des Roches, Équipe de Géophysique Expérimentale, Institut de Physique de Globe de Strasbourg (UMR 7516 CNRS, Université de Strasbourg/EOST), 5 rue René Descartes, Strasbourg cedex 67084, France

ABSTRACT

Background

Geothermal systems are commonly hosted in highly altered and fractured rock. As a result, the relationships between physical properties such as strength and permeability can be complex. Understanding such properties can assist in the optimal utilization of geothermal reservoirs. To resolve this issue, detailed laboratory studies on core samples from active geothermal reservoirs are required. This study details the results of the physical property investigations on Rotokawa Andesite which hosts a significant geothermal reservoir.

Methods

We have characterized the microstructure (microfracture density), porosity, density, permeability, elastic wave velocities, and strength of core from the high-enthalpy Rotokawa Andesite geothermal reservoir under controlled laboratory conditions. We have built empirical relationships from our observations and also used a classical micromechanical model for brittle failure. Further, we compare our results to a Kozeny-Carman permeability model to better constrain the fluid flow behavior of the rocks.

Results

We show that the strength, porosity, elastic moduli, and permeability are greatly influenced by pre-existing fracture occurrence within the andesite. Increasing porosity (or microfracture density) correlates well to a decreasing uniaxial compressive strength, increasing permeability, and a decreasing compressional wave velocity.

Conclusions

Our results indicate that properties readily measurable by borehole geophysical logging (such as porosity and acoustic velocities) can be used to constrain more complex and pertinent properties such as strength and permeability. The relationships that we have provided can then be applied to further understand processes in the Rotokawa reservoir and other reservoirs worldwide.

BACKGROUND

Fractures on multiple scales are the dominant control on fluid flow in most geothermal systems worldwide. Geothermal environments are prone to variable heat fluxes, dynamic fluid flow regimes, and active tectonics which impact the physical and mechanical properties of the reservoir rocks in which they are hosted. The influence of such a dynamic environment can render the host rocks highly altered, fractured, and microstructurally complex. As a result, the empirical correlation of physical properties to yield valuable relationships may not be entirely straightforward. Studies of these properties, and attempts to quantify how they relate to one another in the subsurface, can greatly assist in the optimization and maintenance of geothermal resources (e.g., Gupta and Sukanta [2006]; DiPippo [2008]; Grant and Bixley [2011]).

Here, we detail the results of a systematic physical and mechanical property study on the Rotokawa Andesite; the major reservoir unit within the high-enthalpy Rotokawa Geothermal Field (Krupp and Seward [1987]; Quinao et al. [2013]), located within the Taupo Volcanic Zone (TVZ), North Island, New Zealand (Figure 1). We first examine the texture, mineralogy, petrology, and microstructure. The key physical properties are then explicitly investigated: porosity, density, elastic wave propagation and dynamic elastic moduli, uniaxial compressive strength, static elastic moduli, and permeability. We empirically correlate the microcrack density of the andesite to the measured physical properties. Further, we present empirical relationships of physical properties and classical micromechanical and geometrical models to predict both uniaxial compressive strength and permeability, respectively. Our data is discussed in relation to the Rotokawa Geothermal Field and their applicability to other geothermal resources worldwide.

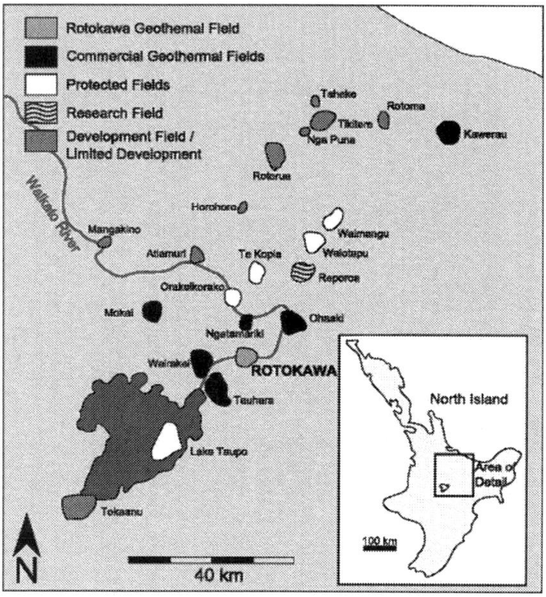

Figure 1: Geothermal fields of the Taupo Volcanic Zone (TVZ), North Island, New Zealand. Core used for this study was sourced from wells drilled in the Rotokawa Geothermal Field. (Adapted from Sewell et al. [2012]).

Previous Studies of Relevance

The study of the core from geothermal systems can yield valuable information to assist their modeling and understanding. For example, Stimac et al. ([2004]) present a study detailing the relationship between permeability and porosity from continuous core from Tiwi geothermal field, Philippines. Their data show that permeability and porosity decrease with depth, with occasional deviations attributed to alteration and compaction. However, the authors are careful to note that their work does not consider the influence of microfractures and their effect on relevant reservoir parameters. Lutz et al. ([2010]) present a case history of the well core from the Desert Peak field (NV, USA) in preparation for the stimulation of an enhanced geothermal system (EGS) by a thorough evaluation of petrological strength and elastic moduli. The results of their study elucidate relationships between clay mineralogy, rock fabric, and permeability increases as a result of mechanical shearing which support proposed hydraulic fracture operations in Well 27-15 at Desert Peak.

The effect of hydrothermal alteration on the physical properties of geothermal core is also a very significant area of research. Hydrothermal alteration can drastically change the elastic wave velocities and permeabilities of rock in both the natural and laboratory environment (Jaya et al. [2010]; Kristinsdóttir et al. [2010]; Pola et al. [2014]). However, coupled studies of physical properties such as porosity, permeability, and strength on geothermal reservoir rocks have not been extensively presented. A detailed study of the impact of a complex microstructure (microfractures and hydrothermal alteration) on the rock physical properties of a geothermal system such as Rotokawa could serve to greatly improve the understanding of reservoir processes at multiple scales.

Geothermal systems are more often than not associated with volcanic systems and are often hosted in rocks sourced from extinct volcanic systems. By proxy, the study of rocks from volcanic edifices can help to boost the understanding of processes within geothermal reservoirs especially with regard to microfractures, which play an essential role in controlling strength, porosity, permeability, elastic wave velocities, and elastic moduli of rocks (Wu et al. [2000]; Guéguen and Schubnel [2003]; Pereira and Arson [2013]; Faoro et al. [2013]; Pola et al. [2014]; Heap et al. [2014]). For example, Vinciguerra et al. ([2005]) studied the influence of thermal stressing on basaltic samples. They show, using elastic wave velocities, that the response of microstructurally variable basalts to thermal stressing can be quite different. While fresh microlitic basalt exhibited severe reductions in P-wave velocity after exposure to 900°C, the P-wave velocity of porphyritic basalt with a pervasive micro crack network did not change.

Similar dependence on the effect of micro fractures on strength (Smith et al. [2009]) and permeability (Nara et al. [2011]) has been investigated, with micro fractures proving to be deleterious to strength and to enhance permeability. Heap et al. ([2014]) showed, for a suite of pervasively fractured and esites, that an increase in porosity from 8 to 29 vol% decreases strength by a factor of 8 and increases permeability by 4 orders of magnitude. David et al. ([1999]) showed that mechanical and thermal micro cracking in granites results in significant changes to permeability and elastic wave velocities. Mechanical micro cracking resulted in the development of P-wave velocity anisotropy, while thermally micro cracked samples showed little P-wave anisotropy. Additionally, permeability was much more varied in mechanically micro cracked rocks than those induced thermally, suggesting that thermal micro cracks develop is otropically. Chaki et al. ([2008]) investigated the role of thermal micro cracking in granites and showed that elastic wave propagation is attenuated by micro cracks and the orientation of these thermal microcracks (with regard to the original microstructure) plays a critical role in the propagation and attenuation of the waves. Faoro et al. ([2013]) provide a model for how micro crack density within an is otropically micro cracked sample can be modeled as a function of aspect ratio and micro crack connectivity. Elastic moduli and elastic wave velocities are strongly influenced by the morphology, distribution, and shape of pore space in rocks and are substantially attenuated by the presence of microcracks (Stanchits et al. [2006] and references therein).

The relationship between porosity and strength has been observed by many authors, with general agreement that as the porosity of a sample (both rock and other engineering materials) increases, the strength decreases (e.g., Al-Harthi et al. [1999]; Li and Aubertin [2003]; Kahraman et al. [2005]; Chang et al. [2006]; Diamantis et al. [2009]; Ju et al. [2013]; Baud et al. [2014]; Heap et al. [2014]). The geometry of the pores also has a significant role in the strength of the materials both intrinsically and with respect to the direction of stress (Luping [1986]). The microstructure of rocks can be changed by increasing the crack damage (by mechanical and/or thermal stresses) as well as hydrothermal alteration (Heap et al. [2009]; Nara et al. [2011]; Pola et al. [2014]); these changes can be observed through the evaluation of destructive and nondestructive physical property measurements (Pola et al. [2012] and references therein; Sousa et al. [2005]). Further, Pola et al. ([2014]) also show that hydrothermal alteration of volcanic rocks can either strengthen or weaken rocks by decreasing or increasing their porosity, respectively.

Geological Significance of the Rotokawa Andesite

The TVZ is a rifted arc associated with the Hikurangi subduction system in which the Pacific plate descends beneath the Australasian plate (Cole [1990]; Wilson et al. [1995]), and hosts active volcanism and multiple associated hydrothermal systems (Bibby et al. [1995]; Rowland and Sibson[2004]; Rowland et al. [2010]). The Rotokawa field is one of these active hydrothermal systems and has been the subject of exploration for mineral resources (sulfur and gold deposits) and, for many years, was the subject of detailed investigation into its use as a commercial geothermal resource (Collar and Browne [1985]; Krupp and Seward [1987]; Hedenquist et al. [1988]). More recently, electricity generation has been realized at Rotokawa following the installations of the Rotokawa I (1997) and Nga Awa Purua (2010) generation stations (Legmann and Sullivan [2003]; Bloomberg et al. [2012]). The more recent of these installations, the Nga Awa Purua power station, hosts the single largest geothermal turbine installation in the world and has a generation capacity >140 MWe which is approximately 3 % of New Zealand's electricity consumption (Horie and Muto[2010]).

The main production zone for the installations at Rotokawa is from that of the Rotokawa Andesite, a series of lavas, pseudo-breccias, and breccias. The movement of fluid through the andesite is predominantly along fracture networks (Rae [2007]; Massiot et al. [2012]). The andesite overlies basement of Miocene greywacke and is capped by a sequence of volcaniclastic and sedimentary units: Reporoa Group, Wairakei Ignimbrite, Waiora Formation, and Huka Falls Formation (Krupp and Seward [1987]; Rae [2007]). The andesite is gray to green and occasionally purple in color, depending on alteration within the reservoir; alteration is less intense in the lavas and more intense in the breccia and pseudo-breccia (Ramirez and Hitchcock [2010]). Production of reservoir fluids is sourced from the Rotokawa Andesite by 12 wells in the central part of the field (Figure 2), and re-injection of spent fluids is done through 5 wells along the southeastern margin of the field (Powell [2011]).

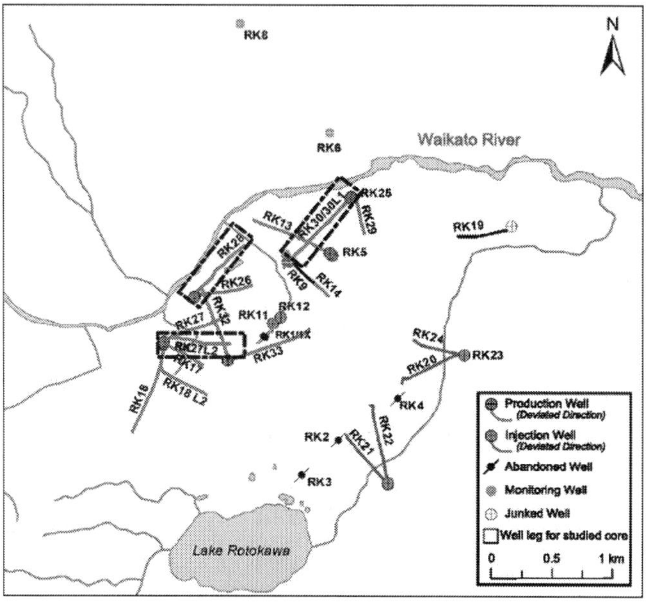

Figure 2: Rotokawa geothermal field and production and injection wells used within the field. Wells RK27L2, RK28, and RK30 were the source of the core used in this study and are outlined with dashed boxes in the figure.

METHODS

Study Source Material

The samples used in this study were sourced from Rotokawa production wells RK27L2, RK28, and RK30 (Figure 2). The measured depths (total borehole length measured from drilling rig floor), orientations, and corrected true vertical depths (TVD) are listed in Table 1. The original cores were approximately 6-m long and 100-mm diameter, and were initially described by the GNS Science Wairakei Research Centre, New Zealand, in a series of internal industry reports detailing the respective production wells from surface to total depth (TD). These reports describe the stratigraphic sequences of the wells and rock types, hydrothermal alteration, and locations of the wells (Rae et al. [2009]; Rae et al. [2010]; Ramirez and Hitchcock [2010]).

Table 1: Detail of core retrieval points from within the Rotokawa Andesite reservoir

Well name	Measured depth of core points (m)	True vertical depth (meters below reference level)	Inclination from vertical (degrees)	Azimuth from north (degrees)
RK28 ST1	2,310 to 2,316	−2,215 to 2,221	21.94	50.91
RK27 L2	2,120 to 2,126	−2,001 to 2,007	27.31	88.11
RK30 L1	2,320 to 2,326	−2,175 to 2,182	20.11	218.47

Measured depths are given as measured from the elevation of the drilling rig floor, true vertical depths are corrected to subsurface elevations, inclination is the deviation from vertical, and azimuth is the orientation of the borehole trajectory.

Siratovich *et al.*

Siratovich *et al.* *Geothermal Energy* 2014 **2**:10, doi:10.1186/s40517-014-0010-4

At the University of Canterbury (UC), the cores were catalogued and cut into workable cylinders approximately 100 mm in length. These smaller sections were over-cored to obtain smaller cylindrical samples 40 mm in diameter and ranging from 80 to 100 mm in length. All samples were machined so that their end faces were flat and parallel in accordance with ISRM standards (Ulusay and Hudson [2007]).

Microstructural Characterization

The strength, porosity, permeability, and acoustic velocities can be significantly influenced by the presence of micro fractures in a sample. Therefore, we deemed it necessary to develop a fundamental understanding of the micro fracture densities in the samples. In order to characterize these features, 10 polished thin sections were prepared from offcuts of the cylinders used for the property characterization described below. The thin sections were prepared perpendicular to the core axis (*X-Y* plane) and, using reflected light thin section photomicrography (at ×40 magnification), were examined for micro fracture densities using the methods suggested by Underwood ([1970])

and further described by Richter and Simmons ([1977]), Wu et al. ([2000]), and Heap et al. ([2014]). In each thin section, an 11 × 11 mm² area was selected, which was subdivided into sections of 1 × 1 mm². The number of cracks that intersected a grid array of parallel and perpendicular lines that were spaced at 0.1 mm was counted. This allowed the calculation of the crack surface area per unit volume according to Equation 1 (Underwood [1970]):

$$\mathrm{Sv} = \pi/2P_{\mathrm{I}} + (2 - \pi/2)P_{\mathrm{II}} \tag{1}$$

Where Sv is the crack surface area per unit volume (in mm²/mm³), P_{I} is the number of perpendicular lines crossed by crack intersections, and P_{II} is the number of parallel lines crossed by crack intersections we also characterized the anisotropy of micro fracture distribution using Equation 2 (Underwood [1970]):

$$\Omega_{23} = P_{\mathrm{I}} - P_{\mathrm{II}}/P_{\mathrm{I}} + (4/\pi - 1)P_{\mathrm{II}} \tag{2}$$

Density and Porosity Measurements

Once the samples were cut and ground flat and parallel, they were washed with water to remove any debris from sample preparation. They were then immersed in distilled water under vacuum of about 100 kPa for 24 h. Samples were taken out of the water and were weighed after their surface water had been removed. The samples were then placed into a laboratory oven at 105°C and dried until a constant mass was observed. Subsequently, they were removed from the oven and held in a dessicator until further characterization was implemented. Sample lengths and diameters were measured to within 0.01 mm. The connected porosity and dry bulk density of the samples were calculated following the methods recommended by Ulusay and Hudson ([2007]).

Characterization of Elastic Wave Velocities and Dynamic Elastic Moduli

The compressional wave (Vp) and shear wave (Vs) velocities and dynamic elastic moduli were measured using a GCTS (Geotechnical Consulting and Testing Systems, Tempe, AZ, USA) Computer Aided Ultrasonic Velocity Testing System (CATS ULT-100) apparatus with axial P- and S-wave piezoelectric transducers (Figure 3). The resonance frequency of the transducers was 900 kHz, pulse acquisition rate was 20 MHz, and 108 waveforms were captured for each sample. The velocities were collected under a constant uniaxial stress of 10 MPa via a Tecnotest servo-controlled 3,000 kN loading frame (Technotest, Modena, Italy) (Figure 3). The stress of 10 MPa was used to ensure a consistent waveform across the specimens and that applied stress was consistent for all measurement cycles. This was determined to be below microcrack closure and opening stress by analyzing the change in axial strain as the sample was loaded to 10 MPa (Eberhardt et al. [1998]). There was no change in axial strain and absence of acoustic emissions (AEs) during the initial loading (Brace et al. [1966]; Martin and Chandler [1994]; Lion et al. [2005]; Nicksiar and Martin [2012]); this ensured a good quality interpretation of the first arrival time of elastic wave pulses. Using these data, we determined the dynamic Poisson's ratio and Young's modulus using Equations 3 and 4 (Guéguen and Palciauskas [1994]), respectively:

$$V_{d} = (Vp^2-2Vs^2)/2(Vp^2- Vs^2) \quad (3)$$

$$E_{d} = (\rho Vs^2(3\,Vp^2-4Vs^2))/(Vp^2- Vs^2) \quad (4)$$

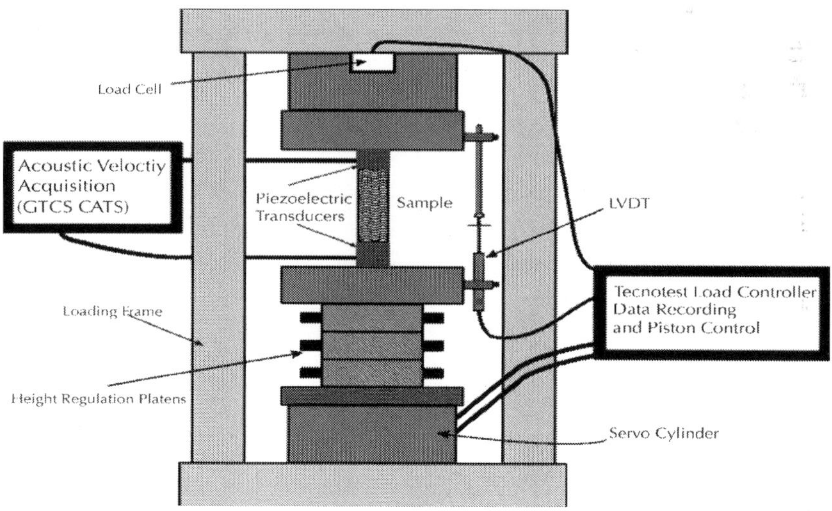

Figure 3: Loading frame set-up for acquisition of elastic velocities at the University of Canterbury. The frame is a Tecnotest 3,000 kN loading frame and

a stress of 10 MPa was applied to each sample cycle to ensure a consistent waveform and quality picking of the first arrival time of the acoustic pulses (note that the figure is not to scale).

Where Vp is compressional wave velocity in meters per second, Vs is shear wave velocity in meters per second, E_d is the dynamic Young's modulus in pascal, V_d is the dynamic Poisson's ratio, and ρ is density in kilograms per cubic meter. Our physical property characterizations are summarized in Table 2. In addition to the determination of the elastic moduli from our elastic wave velocity measurements, we also utilized the method of Martínez-Martínez et al. ([2011]) to characterize the spatial attenuation of the compressional waveform anisotropy (Equation 5):

$$\alpha_s = 20 \log (A_e / A_{max}) / L \tag{5}$$

where α_s is spatial attenuation in decibels per centimeter, A_e is the maximum amplitude emitted by the piezoelectric crystal, A_{max} is the recorded maximum amplitude of the pulse after passing through the sample, and L is the length of the sample in meters.

Table 2: Results of quantitative microstructural characterization

Sample name	Crack density for intercepts parallel to orientation axis*P*|| (mm−1)	Crack density for intercepts perpendicular to orientation axis*P* | (mm−1)	Crack area per unit volume Sv (mm2/mm3)	Anisotropy factorΩ2,3	Connected porosity (vol %)
27_21_0B	13.73	13.53	13.06	0.01	14.91
28_10_5A	4.10	4.41	8.33	0.06	7.47
27_20_4_B	1.83	2.06	3.77	0.08	4.37
27_3_3B	4.77	4.77	9.55	0.01	9.81
30_22_4A	2.77	2.62	5.48	0.01	6.49
28_10_9B	3.84	3.90	7.71	0.03	7.42
28_12_1	4.83	4.65	9.59	0.02	7.89
30_21_1B	2.93	2.91	5.85	0.01	7.47
27_21_3A	5.80	5.13	11.31	0.09	16.3
28_10_6C	2.97	2.87	6.38	0.05	5.97

As discussed in the section on microstructural characterization, crack densities were calculated on thin section samples to ascertain crack areas per unit volume using the optical microscope method (Underwood [1970]).

Siratovich *et al.*

Siratovich *et al. Geothermal Energy* 2014 2:10, doi:10.1186/s40517-014-0010-4

Uniaxial Compressive Strength Testing and Static Elastic Moduli

Uniaxial compressive strength (UCS) was determined using a Technotest 3,000 kN, servo-controlled loading frame (Figure 4). Four strain gauges (20-mm strain gauges with a gauge factor of 2.12 supplied by Tokyo Sokki Kenkyujo Co. Ltd. (TML) Shinagawa-ku, Tokyo, Japan) were glued onto each sample. Two vertical gauges measured axial strain and two laterally oriented gauges measured radial strain; care was taken to ensure that the strain gauges were perpendicular to their respective axes of deformation. The specimens were deformed at a constant strain rate of 1.0×10^{-5} s^{-1} (controlled by linear variable differential transformer, LVDT) at ambient laboratory temperature and humidity conditions. During experimentation, AE output was monitored using Physical Acoustics Corporation MISTRAS' AE node acquisition system (Princeton Jct, NJ, USA). Two physical acoustics WSα AE transducers (100 to 900 kHz operating frequency) were attached to the samples at the top and base, and hit counts, waveforms, energy, and amplitude of the received signals were recorded during sample deformation. AE monitoring was used during deformation as a proxy for microcracking as AEs are generated by the release of energy from a material during the propagation and nucleation of microcracks (Eberhardt et al. [1998]; [Diederichs et al. 2004]). We utilized arbitrary AE energy units (the area under the received waveform signal) for comparison of AE activity across the datasets. Once stress-strain curves were obtained and AE data are processed, we calculated the static elastic moduli for each specimen utilizing Equations 6 and 7 with the tangent deformation modulus at 50 % of the maximum peak stress (Ulusay and Hudson [2007]). In addition, we selected portions of the stress-strain sequence to identify crack closure, crack initiation, unstable crack propagation, and, ultimately, crack coalescence and sample failure (Martin [1993]; Eberhardt et al. [1998]; Takarli et al. [2008]; Heap and Faulkner [2008]):

$$E_s = (\Delta\sigma a / \Delta\varepsilon a) \tag{6}$$

$$\nu_s = -(\Delta\varepsilon r / \Delta\varepsilon a) \tag{7}$$

where E_s is the static Young's modulus (Pa), ν_s is the static Poisson's Ratio, σa is the differential axial stress (Pa), εa is the axial strain, and εr is the radial strain.

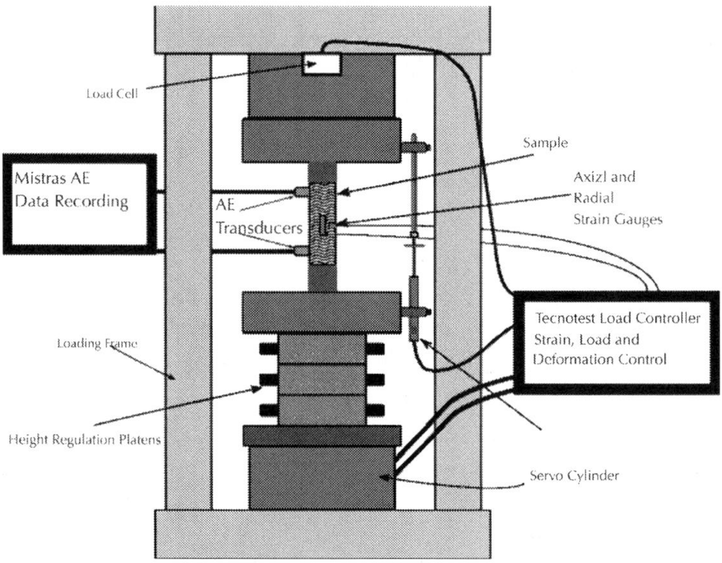

Figure 4: Loading frame set-up for determining uniaxial compressive strength (UCS) at University of Canterbury. Testing was carried out using TML strain gauges and MISTRAS acoustic emission monitoring equipment. Loading during axial differential stressing was achieved through a constant strain rate of 1.0×10^{-5} s^{-1} (not to scale).

Permeability Measurements

Gas (argon) permeability measurements were made at the Laboratoire de Déformation des Roches, Université de Strasbourg (France). The 40-

mm diameter specimens were over-cored to a diameter of 20 mm and cut and ground flat and parallel to a nominal length of 40 mm. The new samples were then re-evaluated by the triple-weight method to obtain porosity via the Archimedes' method (Ulusay and Hudson [2007]) and oven-dried under vacuum at 40°C until no change in sample mass was observed. The samples were then jacketed with viton sleeves, placed between two steel end-caps and lowered into the pressure vessel (Figure 5). A confining pressure of 2 MPa was applied to the sample (provided by distilled water), and permeability measured using the transient method (or pulse-decay method). For the permeability measurements, an initial differential pore pressure was applied to the sample, the upstream inlet was then closed, and the pore pressure decay monitored over time. The downstream fluid pressure (P_{down}) was the ambient atmospheric pressure, and the maximum upstream fluid pressure (P_{up}) was set so the pressure differential was 0.5 MPa. Permeability was then calculated using Equation 8 (after Brace et al.[1968]):

$$k_{gas} = (2\eta L/A)\left(V_{up}/(P_{up})^2-(P_{down})^2\right)(\Delta P_{up}/\Delta t) \tag{8}$$

where k_{gas} is the gas permeability, η is the viscosity of the pore fluid, A is the cross-sectional area of the sample, V_{up} is the volume of the upstream pore pressure circuit (approximately 7 cm^3), P_{up} is the upstream pore pressure, P_{down} is the downstream pore pressure, and t is the time. By plotting ΔP_{up} as a function of time, the local slope of the curve is computed to determine the temporal variation of the permeability k_{gas}. To check whether our data should be corrected for Klinkenberg's 'slip flow' (Klinkenberg [1941]), we plotted the measured gas permeability as a function of the inverse of the mean pore fluid pressure, P_{mean}. For the transient method, sinceP_{down} is constant, the decay of P_{up} through time corresponds to the decay of the mean pore pressure P_{mean}. We found that, in all cases, the Klinkenberg correction should be applied:

$$k_{true} = k_{gas}(1 + b/ P_{mean}) \tag{9}$$

where k_{true} is the true gas permeability, b is Klinkenberg slip factor, and P_{mean} is the mean pore fluid pressure.

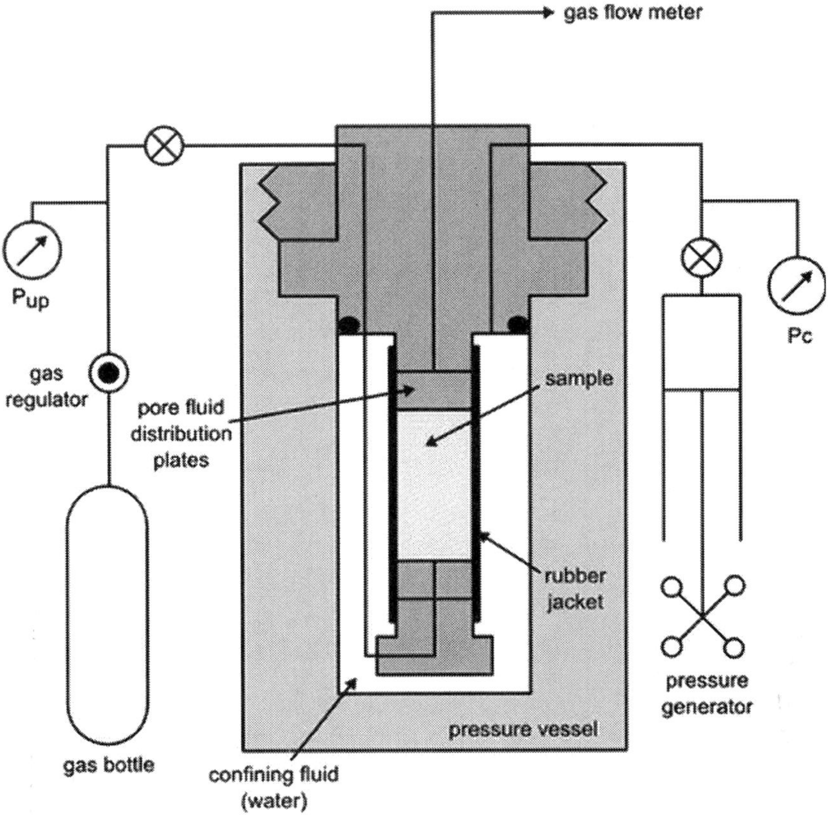

Figure 5: Gas permeameter used to measure permeability at University of Strasbourg (not to scale). Confining pressure of 2 MPa was applied using distilled water as the pressurizing media, and pore fluid was dry argon gas supplied at 1.5 MPa for a net effective pressure of 0.5 MPa.

RESULTS

In the following section, we present our data and observations on petrology, microstructure (quantitative microfracture analysis), macrostructure (bulk density, porosity, acoustic wave velocities, and dynamic moduli), strength relations (by UCS testing), and finally the ability of the rock to transmit fluid (permeability) of the Rotokawa Andesite.

Petrology

The Rotokawa Andesite shows moderate to intense hydrothermal alteration with the groundmass and phenocrysts showing replacement of original mineralogy. Fractures and occasional veins of quartz, calcite, anhydrite, and epidote occur, and amygdales within the sample are often filled with chlorite, calcite, hematite, pyrite, and chalcedony, often with quartz rims (Figure 6). Alteration is pervasive with the original mineral assemblages typically replaced by secondary hydrothermal alteration species, with some specimens showing very little original mineralogical texture. Plagioclase feldspars have been altered to albite, adularia, occasional calcite, and rare pyrite, and ferromagnesian minerals have been replaced by chlorite, quartz, calcite, and occasional epidote. Microfractured phenocrysts (Figures 6 and 7) are abundant and many relict phenocrysts retain original texture are but replaced by secondary mineralization. The average phenocryst size is 0.5 to 1 mm with occasional plagioclase near 1.5 to 2 mm; amygdales also range from 1 to 1.5 mm in size. The alteration chemistry of the samples indicates that this portion of the reservoir is dominated by chlorite/epidote alteration. The degree of alteration is relatively consistent across the core we have sampled with most primary mineralogies replaced by secondary alteration products. Microfracture mineralization indicates that these networks may have been conductive pathways for fluid migration (i.e., the presence of chlorite clays, adularization of plagioclase, calcite, and quartz rimming of fractured matrix); we typically observe chlorite, calcite, and quartz as alteration mineralogies with occasional epidote centers within the fractures. Backscatter scanning electron microscopy (SEM) was utilized to further reiterate the complex interaction of fractures and vesicles in the specimens (Figure 7). At several different magnifications, we see an abundance of microfractures in the samples as well as a clear depiction of the complex alteration mineralogy displayed by the andesite.

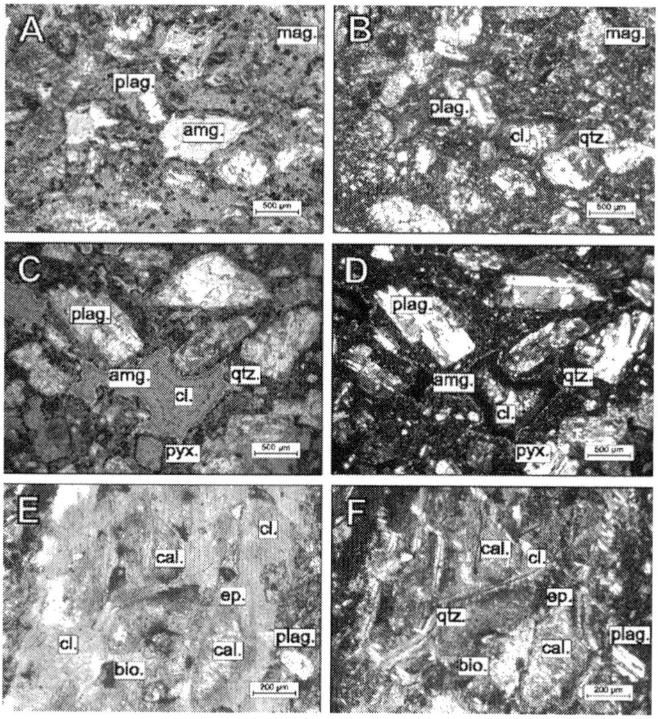

Figure 6: Thin section photomicrographs of the Rotokawa Andesite. (A) Plane-polarized light of RK28 2310.6C andesitic pseudo-breccia with plagioclase laths (plag.), groundmass is altered plagioclase, abundant magnetite (mag.), and amygdale (amg.). (B) Cross-polarized light of RK28 2310.6C clearly shows alteration fabrics of the brecciated andesite with plagioclase (plag.) and amygdales filled with chlorite (cl.) and rimmed by quartz (qtz.). (C) Plane-polarized light view of RK27_L2 2121.4A showing andesitic breccia with plagioclase with slight adularia alteration (plag.) and amygdale (amg.) filled with chlorite (cl.), quartz (qtz.), and highly altered pyroxene (pyx.) in lower portion of image. (D) Cross-polarized light view of RK27_L2 2121.4A shows chlorite infill of a large amygdale (amg.) in the center of the photomicrograph and quartz rim (qtz.), plagioclase (plag.), and highly altered pyroxene (pyx.). (E) Plane-polarized light view of RK30 2322.4A shows highly altered and microfractured plagioclase phenocryst with intense alteration and replacement by chlorite (cl.), epidote (ep.), calcite (cal), possible biotite (bio.), and small plagioclase showing evidence of adularia alteration (plag.). (F) Cross-polarized light of RK30 2322.4A illustrates microfracture network and veining with alteration products of quartz (qtz.), epidote (ep.), biotite (bio.), calcite (cal.), and plagioclase (plag.).

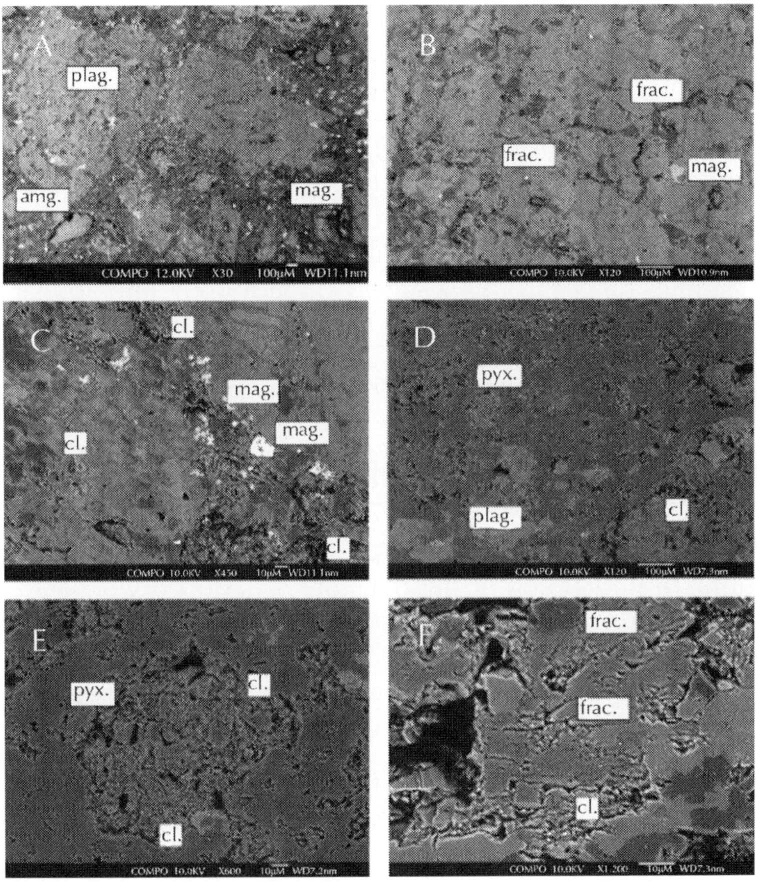

Figure 7: Backscattered scanning electron microscope photomicrographs of Rotokawa Andesite. Samples A to C are from RK28 at 2,310.6 m depth and D to F are from RK27 at 2,121.1 m depth.(A) Andesite with abundant ferromagnesian minerals (mag.), altered amygdale (amg.), and highly scattered magnetites (mag. bright hues).(B) Detail of a fractured plagioclase phenocryst showing the microfractured texture (frac.) and occasional magnetite (mag.). (C)Detail of a fracture infill showing chloritization (cl.) and abundant ferromagnesian minerals (mag.). (D) Groundmass of RK27 sample; pervasive fracturing is not apparent at this magnification but the porous network is quite apparent with pyroxene (pyx.), chlorite (cl.), and plagioclase (plag.). (E) Replacement mineralogy of likely pyroxene phenocryst (pyx.) showing abundant chloritization (cl.) and dissolution textures. (F) Detail of relict pyroxene and abundant chlorite (cl.) with abundant microfractures (cl.) apparent in the sample mass.

Quantitative Two-Dimensional Microstructural Analysis

We evaluated the microfracture density of 10 specimens as a function of crack surface area per unit volume (Table 2). These samples were selected to represent the range of connected porosities observed within the sample set. We found that the crack area per unit volume in our samples ranges from 3.77 to 13.06 mm^2/mm^3 and appears to be independent of the alteration and mineralogy of the specimens. The calculated anisotropy factor ($\Omega_{2,3}$), indicates that the microcracks are isotropic (Table 2).

Porosity and Bulk Density

Bulk density decreases as connected porosity increases, as expected for samples of similar composition (Figure 8A). Bulk dry densities of the samples range from 2.29 to 2.65 g/cm^3, with a mean value of 2.49 g/cm^3 The connected porosities range from 4.37 to 16.3 vol%, with a mean value of 8.44 vol%. While there is some variation in the distribution of pores/vesicles in the samples, we observe that the microcrack density exerts an important control on the porosity and density, as illustrated by the correlation between crack area per unit volume and connected porosity presented as Figure 8B.

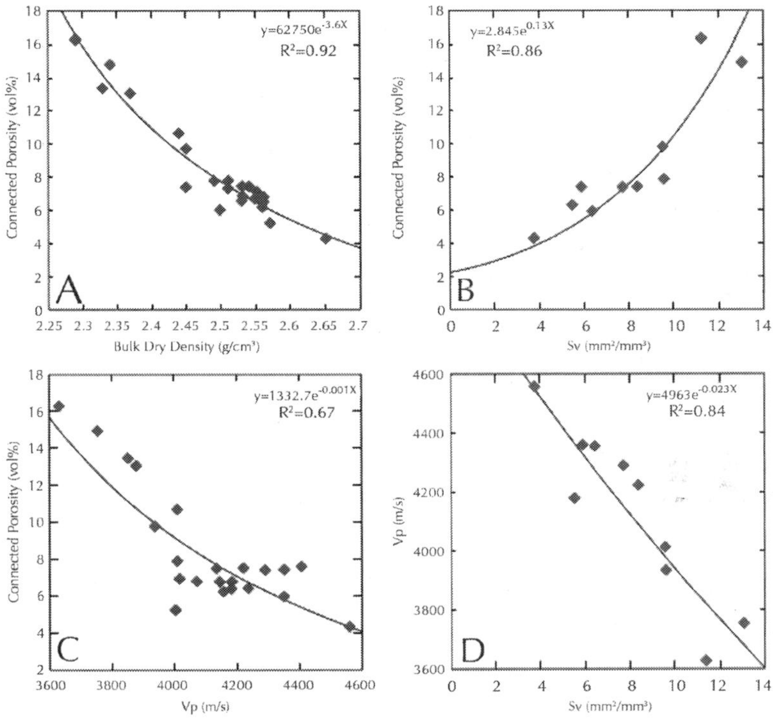

Figure 8: Relationships of porosity, density, crack area, and compressional wave velocity for Rotokawa Andesite. (A) Connected porosity versus dry bulk density for Rotokawa Andesite calculated using the dual weight method of (Ulusay and Hudson [2007]). (B) Crack area per unit volume (Sv) plotted versus connected porosity for samples representing the range of values measured within the measured dataset.(C) Connected porosity (vol%) using the dual weight method versus axial compressional wave velocity (Vp) as measured under a stress of 10 MPa which was determined to be below the crack closure stress (see text for further detail). (D) Crack area per unit volume (Sv) versus compressional wave velocity (Vp).

Ultrasonic Wave Velocities, Dynamic Elastic Moduli, and Spatial Attenuation

Measurements made on dry samples under ambient (pressure and temperature) conditions yielded axial P-wave velocities from 3,627 to 4,556 m/s with a mean value of 4,106 m/s, and axial S-wave velocities between 2,160 to 2,752 m/s with a mean value of 2,510 m/s (Table 3). Porosity and P-wave velocities show moderate correlation: samples with higher porosities have slower elastic wave velocities, as seen in Figure 8C. The crack area per unit volume also correlates well with P-wave velocity (Figure 8D). The axial spatial attenuation for the andesites ranges from 8.39 to 28.74 dB/cm (Figure 9). Figure 9 shows that there is no clear trend between spatial attenuation and P-wave velocity. Dynamic Poisson's ratio and dynamic Young's modulus were in the range of 0.13 to 0.23 and 24.6 to 45.9 GPa, respectively (Table 3).

Table 3: Physical property measurements of 22 samples used in destructive testing of Rotokawa Andesite

Sample source_ well sample name	Bulk dry density (g/cm3)	Connected porosity (vol%)	Vp (m/s)	Vs (m/s)	Spatial attenuation (dB/cm)	UCS (MPa)	Static Young's modulus (GPa)	Dynamic Young's modulus (GPa)	Static Poisson's ratio	Dynamic Poisson's ratio
RK_27_ L2_21.5B	2.44	10.72	4,005	2,443	14.63	85.99	19.9	35.1	0.24	0.2
RK_27_ L2_21.8A	2.33	13.49	3,850	2,363	16.47	79.91	25.2	31.2	0.26	0.2
RK_27_ L2_23.2A	2.56	6.61	4,182	2,490	22.48	105.26	31.2	38.9	0.19	0.23
RK_27_ L2_20.4B	2.65	4.37	4,556	2,752	22.53	211.05	37.7	45.9	0.25	0.21
RK_27_ L2_21.1C	2.37	13.1	3,877	2,405	23.6	69.53	21.5	32.5	0.18	0.19
RK_27_L2_3.3B	2.45	9.81	3,937	2,331	23.51	95.78	32.4	29.9	0.13	0.18
RK_27_ L2_21.0B	2.34	14.91	3,752	2,337	8.4	60.13	28.1	30.6	0.12	0.17
RK_27_ L2_21.3A	2.29	16.3	3,627	2,160	15.13	70.57	30.4	24.6	0.16	0.17
RK_28_10.6C	2.5	5.97	4,350	2,652	18.27	146.2	43.7	42.4	0.27	0.2
RK_28_10.8C	2.53	6.72	4,147	2,537	11.67	109.91	27.2	39.1	0.34	0.2
RK_28_10.9B	2.51	7.42	4,285	2,615	14.67	137.31	32.4	41.5	0.2	0.2
RK_28_13.2A	2.55	6.97	4,013	2,531	18.53	146.21	38.3	37.2	0.24	0.19
RK_28_10.5A	2.45	7.47	4,220	2,578	14.83	130.71	27.4	38.8	0.25	0.21
RK_28_11.5A	2.51	7.62	4,403	2,555	10.06	152.76	35.6	37.4	0.27	0.13
RK_28_12.1	2.49	7.89	4,010	2,460	12.23	115.01	29.3	36.2	0.22	0.14

RK_30_20.4A	2.57	5.3	4,002	2,495	10.39	140.97	33.6	36.9	0.09	0.15
RK_30_21.0A	2.55	6.84	4,070	2,508	11.17	126.53	26.4	39.8	0.14	0.15
RK_30_21.1B	2.54	7.47	4,352	2,659	28.75	157.93	25.8	43.2	0.17	0.2
RK_30_21.7B	2.56	6.28	4,154	2,588	19.91	162.71	28.3	41.1	0.22	0.18
RK_30_22.3B	2.53	7.51	4,133	2,550	21.72	137.97	31.5	39.2	0.23	0.19
RK_30_22.4A	2.56	6.49	4,181	2,582	20.27	148.44	34.4	40.7	0.18	0.19
RK_30_22.5B	2.56	6.41	4,236	2,628	23.43	150.71	33.6	42.1	0.09	0.16
Mean	2.49	8.44	4,106	2,510	17.39	124.62	30.6	37.5	0.2	0.18
Standard deviation	0.09	3.23	221	133	5.512	37.01	5.5	5.1	0.06	0.03

Siratovich *et al.*

Siratovich *et al. Geothermal Energy* 2014 **2**:10, doi: 10.1186/s40517-014-0010-4

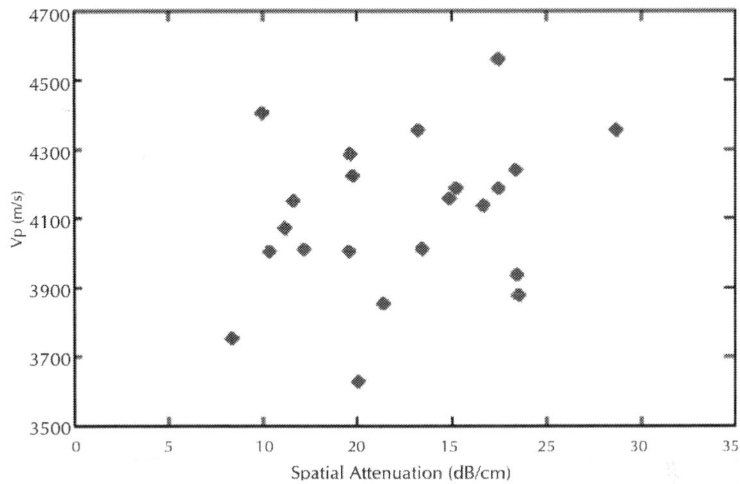

Figure 9: Spatial attenuation (**α**s) of axial compressional P-wave velocity of Rotokawa Andesite. Attenuation was calculated by the method suggested by Martínez-Martínez et al. ([2011]) utilizing transmission of the ultrasonic wave and maximum attenuation of the waveform plotted versus axial compressional wave velocity (Vp) obtained under an axial stress of 10 MPa.

Uniaxial Compressive Strength and Static Elastic Moduli

In order to characterize the mechanical behavior of the Rotokawa Andesite, the 22 samples of Table 3 were loaded in compression to failure. The dataset shows a large range of UCS (as observed for the other physical properties), from 60 to 211 MPa. The stress-strain behavior of the andesites is very similar across the range of strengths, as shown in Figure 10, which reports curves that best represent the dataset and behavior of the Rotokawa Andesite under uniaxial compression. All specimens in the dataset show brittle behavior as evidenced by the stress-strain relationships and bolstered by analysis of the AE activity. An increase of AE between dilatancy (σ_{cd}) and failure is a benchmark of brittle failure (e.g., Brace and Bombolakis [1963]; Rutter[1986]; Ashby and Sammis [1990]; Heap and Faulkner [2008]), as seen in Figure 10. Weaker specimens showed lower overall AE energy output than higher strength specimens. Static Young's moduli range between 19.9 and 43.7 GPa and static Poisson's ratio between 0.09 and 0.34. Our data shows that as porosity and crack surface area increases, the UCS of the rock decreases (Figure 11A,B). Further, the UCS increases as axial P-wave velocity increases (Figure 11C).

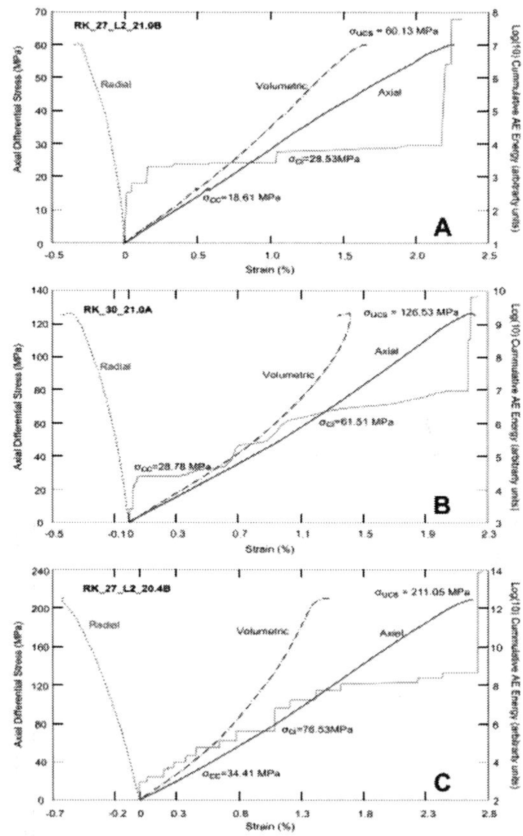

Figure 10: Stress–strain behavior of the Rotokawa Andesite.Samples were subject to constant strain rate loading (1×10^{-5}/s) and monitored for associated arbitrary acoustic emission energy output. All samples in this study display brittle failure. (A) Samples with low UCS generally develop a single fracture plane. (B) Samples near mean UCS develop several fracture planes. (C) Samples with very high UCS showed explosive, catastrophic failure into several large and small pieces with no distinct failure plane.

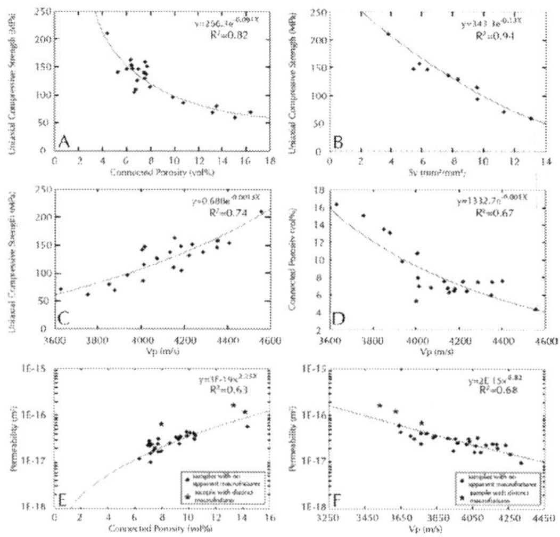

Figure 11: Key empirical relationships of the Rotokawa Andesite. (A) Connected porosity (vol%) plotted versus uniaxial compressive strength (MPa) for Rotokawa Andesite at ambient pressures and temperatures. (B) Crack area per unit volume (Sv) as measured from reflected light thin sections (method recommended by Underwood [1970]) plotted versus uniaxial compressive strength (UCS) for the Rotokawa Andesite. (C) Axial P-wave velocity (Vp) as measured at ambient temperatures under an axial load of 10 MPa versus uniaxial compressive strength values for the Rotokawa Andesite at ambient temperature. (D) Axial compressional wave velocity (Vp) plotted versus connected porosity (vol%) as measured by the dual weight method (Ulusay and Hudson [2007]). (E) Semi-log plot of connected porosity measured by the triple-weight method (Ulusay and Hudson [2007]) versus argon gas permeability with effective pressure of 0.5 MPa. (F) Semi-log plot of axial compressional wave velocity (Vp) plotted versus argon gas permeability.

Permeability

Our argon permeability measurements show that for the tested samples, permeability ranges from 9.82×10^{-18} m^2 to 1.66×10^{-16} m^2 (Table 4). The results show a trend of increasing permeability with increasing porosity (Figure 11E). We observe that three of the samples contain macrofractures (black stars on Figure 11E,F) and have higher permeabilities that slightly deviate from the trend of the dataset. We also note that as permeability increases, the axial compressional wave velocity decreases (Figure 11F); further, those samples with distinct macrofractures show lower compressional velocities and higher permeabilities when compared with samples of similar porosity.

Table 4: Results of density, porosity, argon permeability, and acoustic velocity measurements for Rotokawa Andesite

Sample source: well number, depth, name	Bulk dry density (g/cm3)	Connected porosity (vol%)	Argon permeability (m2)	Axial P-wave velocity(m/s)
RK_27_L2_2120.4A_1	2.55	7.24	1.78E−17	3,943
RK_27_L2_2120.4A_2	2.55	7.06	9.82E−18	4,318
RK_27_L2_2121.1A	2.34	13.97	1.25E−16	3,621
RK_27_L2_2123.3A_1	2.48	9.22	3.32E−17	3,911
RK_27_L2_2123.3A_2	2.46	10.25	4.13E−17	3,793
RK_27_L2_2123.7A_1	2.52	9.73	3.69E−17	3,840
RK_27_L2_2123.7A_2	2.50	10.35	3.72E−17	3,704
RK_27_L2_2124.1B_1	2.54	7.63	3.24E−17	3,954
RK_27_L2_2124.1B_2	2.56	6.93	2.81E−17	4,032
RK_27_L2_2121.0A_1	2.42	13.14	1.66E−16	3,532
RK_27_L2_2121.0A_2	2.33	14.23	6.09E−17	3,642
RK_28_2310.3A_1	2.46	9.94	4.26E−17	3,760
RK_28_2310.3A_2	2.48	9.70	4.44E−17	3,650
RK_28_2310.8A_1	2.50	8.95	3.45E−17	3,830
RK_28_2310.8A_2	2.50	9.22	3.43E−17	4,076
RK_28_2310.9C_1	2.46	10.30	3.21E−17	3,725
RK_28_2310.9C_2	2.53	7.64	1.65E−17	4,096
RK_28_2311.1B_1	2.55	7.04	1.53E−17	4,259
RK_28_2311.1B_2	2.56	7.35	2.05E−17	4,190
RK_28_2311.3B_1	2.50	9.10	3.21E−17	4,026
RK_28_2311.3B_2	2.49	9.17	2.50E−17	4,183
RK_30_2320.6A_1	2.53	7.90	7.10E−17	3,765
RK_30_2320.6A_2	2.56	7.79	1.70E−17	4,061
RK_30_2321.0B_1	2.50	8.24	2.50E−17	3,762
RK_30_2321.0B_2	2.54	8.47	2.50E−17	4,164
RK_30_2321.2B_1	2.58	7.11	2.38E−17	4,235
RK_30_2321.2B_2	2.55	6.98	2.37E−17	4,122
RK_30_2321.5A	2.57	6.92	2.60E−17	3,965
RK_30_2322.3A_1	2.54	7.61	2.44E−17	4,016
RK_30_2322.3A_2	2.57	6.83	2.27E−17	4,069

Siratovich *et al.*

Siratovich *et al.* *Geothermal Energy* 2014 **2**:10, doi:10.1186/s40517-014-0010-4

DISCUSSION

Micromechanical Interpretation

We have shown that the Rotokawa Andesite contains a pervasive network of isotropic microcracks. Due to their isotropic distribution, the majority of these microcracks are consistent with the results of thermal stressing (Fredrich and Wong [1986]; Reuschlé et al. [2006]; Wang et al. [1989]; David et al. [1999]; Heap et al. [2014]). Indeed, the Rotokawa Andesite has experienced several cycles of heating and cooling: the initial eruption of the andesite, burial in a faulted graben, hydrothermal alteration, and the eventual exhumation during core recovery (Rae [2007]; Lim et al. [2012]). Our microstructural analysis has highlighted that the pervasive microcracking appears independent of lithology, original mineralogy, and secondary (hydrothermal alteration) mineralogy.

The intense microcracking in our samples has shown to be a significant factor in all of the measured physical properties. First, microcracking has greatly reduced the propagation velocity of elastic waves through the andesite. We see a clear correlation of crack area per unit volume (Sv) to the observed compressional wave velocities (Figure 8D) and interpret this to be attenuation of the compressional wave through the cracked intracrystalline and intercrystalline boundaries that are abundant in the andesite (e.g., Figures 3 and 4). Many authors (e.g., Vinciguerra et al. [2005]; Keshavarz et al. [2010]; Blake et al. [2012]; Heap et al. [2014]) have also shown that the elastic wave velocities can be highly attenuated by the presence of microcracks.

Second, the crack surface area and UCS have yielded an excellent correlation (Figure 11B). As noted by Walsh ([1965a], [b]), David et al. ([1999]), and Chaki et al. ([2008]), the density of the cracks within a specimen is critical in dictating its strength. The development of microcracks during uniaxial compression, and the coalescence of these cracks (newly formed and pre-existing), leads to the failure of the sample (Brace et al. [1966]; Bieniawski [1967]). In samples that already show relatively high crack densities, less energy is required to coalesce existing cracks and thus they are inherently weaker (David et al. [1999]; Ferrero and Marini [2001]; Keshavarz et al. [2010]). By utilizing AE monitoring during our UCS testing, we observe that fewer events occur during uniaxial compression in weaker samples than those with higher strength (Figure 10), indicating that there are far more pre-existing cracks in the weaker samples (Hardy [1981]; Eberhardt et al.[1998]; Nicksiar and Martin [2012]). Thus, the presence of pre-existing microcracks in the Rotokawa Andesite is shown to exert a strong control on their uniaxial compressive strength.

Permeability is one of the most important properties of a geothermal system. In this study, we have seen that porosity (and bulk sample density) and strength are related to the extent of the microcracking in the andesite. We did not measure the crack surface area in the samples used for our permeability measurements (the samples will be used for future studies; calculating crack surface area required destructive thin section preparation). However, we can, by proxy, assume a correlation between permeability and the extent of the microfracture network. We show that there is a clear inverse relationship between the sample's permeability and P-wave velocity such that as permeability increases, compressional wave velocity decreases (Figure 11F). These results are consistent with the many investigations have shown a clear link between reduced elastic wave velocities and increased permeability (David et al. [1999]; Vinciguerra et al. [2005]; Chaki et al.[2008]; Nara et al. [2011]; Faoro et al. [2013]; Heap et al. [2014]). While we have not measured the relationship of crack density to permeability directly in our dataset, we show that Sv and Vp are inversely related (Figure 8D), and a similar relationship exists between Vp and permeability. Therefore, we can infer that those samples with higher crack surface areas will be inherently more permeable.

Key Empirical Relationships

In this section, we present relationships of singular variables that could be readily and easily measured either using photomicrography or geophysical logging tools and their correlation to more complicated and pertinent physical properties. All of these parameters are singularly measurable variables that do not rely on complex formulae for their derivation (such as dynamic Young's Modulus or Poisson's ratio) and so have been selected to be the key relationships that we present with relevance to the Rotokawa Andesite.

Porosity and UCS

An exponential correlation between sample porosity and UCS exists (Figure 11A). Such correlations have been utilized by several authors (e.g., Vernik et al. [1993]; Li and Aubertin[2003]; Palchik and Hatzor [2002]; Kahraman et al. [2005]; Chang et al. [2006]; Palchik [2013]; Pola et al. [2014]) for a variety of clastic and volcanic rocks and concrete materials. These authors present empirical fits for the correlation of physical properties versus UCS and show a wide range of correlation within their respective datasets with R^2 values from near 0.6 to as high as 0.95. We propose that our empirical fit between porosity and UCS (an exponential fit with a correlation factor of 0.82, Figure 11A) can provide useful estimations of the strength of the reservoir rocks within the Rotokawa Andesite reservoir. By utilizing estimations of UCS derived from the correlation of porosity, the minimum strength of the rocks can be applied to important engineering issues such as wellbore stability (Chang et al. [2006]; Schöpfer et al. [2009]).

Vp and UCS

There is an exponential correlation between strength and Vp with an R^2 value of 0.74 (Figure 11C). As noted by Kahraman ([2001]), the relationship between Vp and UCS is generally nonlinear and the higher the strength of the material, the more scattered the data points. Heap et al. ([2014]) came to similar conclusions following measurements on andesitic rocks from Volcán de Colima (Mexico). In our study, there is an increasing trend of strength with increasing Vp but, as shown in Figure 9, there is a high degree of spatial anisotropy with respect to Vp such that a robust correlation of strength to elastic wave velocity is difficult to obtain. However, Vp is a widely utilized logging tool in borehole geophysics (Chang et al. [2006]), and using the correlation that we have obtained, a minimum strength criteria could be established from the response of the logging tool. This is an important correlation as geophysical logging is much easier, faster, and more efficient than cutting spot cores (as the core for this study was obtained), and so the development of empirical correlations to constrain strength such as that seen in Figure 11B can help mitigate risk and reduce the cost associated with geothermal drilling programs.

Vp and Porosity

Correlations between Vp and porosity show an increasing trend of porosity with decreasing Vp (Figure 11D, also observed by Al-Harthi et al. [1999]; Rajabzadeh et al. [2011]; Tugrul and Gurpinar [1997]; Heap et al. [2014]). This can be attributed to both the pore structure distribution and the degree of microcracking within the andesites. It is clear from microstructural analysis (using both optical and scanning electron microscope analyses) that a large proportion of the porosity in the Rotokawa Andesite is likely to be composed of (macro- and mesoscale) fractures and microcracks (e.g., Figures 6 and 7).

An explanation for the variation and wide distribution of the elastic wave velocity data for samples with similar porosities (specifically with regard to those data that range from 4,000 to 4,400 m/s) is that there must be a variable pore (vug/vesicle) content or hydrothermal alteration between the samples. The presence of pores will greatly augment the porosity (due to their aspect ratio) but will have comparatively little influence, compared to the microcracks, on the P-wave velocity. The application of our exponential relationship (Figure 11D) can give a rough approximation for seismic velocities derived from connected porosity, or *vice versa*. This may be useful during the drilling of additional wells at Rotokawa where porosity can be measured at the wellsite and yield a rough approximation for P-wave velocities and, as such, tie back to our empirical correlations of strength (Figure 11C).

Permeability and Porosity

Our permeability and porosity data show that there is a clear trend of increasing porosity with increased permeability for the Rotokawa Andesite (Figure 11E), a common observation in multiple lithologies (e.g., Heard and Page [1982]; Géraud [1994]; Stimac et al. [2004]; Chaki et al. [2008]; Watanabe et al. [2008]; Heap et al. [2014]). We observe that our relationship between porosity and permeability can be described by a power law correlation and is consistent with the Kozeny-Carman relation (Guéguen and Palciauskas [1994], see the 'Application of micromechanical and geometrical permeability models' section). The dependence of permeability on porosity is generally explained by the assumption that a more connected pore space (cracks and pores) provides more efficient pathways for fluid migration (e.g., Costa [2006]; Chaki et al. [2008]). We do however need to consider those data points that have a very similar value of permeability (approximately 3.2×10^{-17} m^2, Table 4), with a porosity range of 7.6 to 10.3 vol% that indicate that there is variability of the samples with respect to permeability that may be reflected in the tortuosity of the porous network. This is consistent with the findings of Bernard et al. ([2007]) and Heap et al. ([2014]) such that the permeability in volcanic rocks is highly dependent upon connectivity of the microstructure.

With respect to microstructure, we have shown that the porosity is very closely linked to crack surface area (Figure 8D) and, thus, that increasing crack density corresponds to a sample with a higher permeability. The three samples that lie slightly outside the trend of the dataset display distinct mesofractures (black stars in Figure 11E,F) and that these mesofractures greatly enhance the permeability of the samples without significantly increasing their porosity. These specimens show higher than average permeability for their porosity, which supports the conclusions of Stimac et al. ([2008]) that meso- and macrofractures are critical in controlling the permeability of geothermal reservoir systems. On the large scale, macrofractures are necessary for fluid production from geothermal reservoirs, but the microstructural characteristics of the host rocks cannot be neglected when considering fluid flow, storage capacity, and total permeability of the reservoir (Jafari and Babadagli [2011]).

The robust relationship between porosity and permeability has wider-scale reservoir applications where the need to understand reservoir rock permeability (the mass itself, not those portions with highly macroscopic fractures e.g., Massiot et al. [2012]) is important for reservoir forecasting and modeling. Measurements of porosity can then yield a good approximation of the permeability of the intact reservoir rock at Rotokawa through our power law correlation (Figure 11E). However, we urge caution if the porosity falls outside our measured range. As porosity is a readily measureable property by geophysical logging tools (Ellis and Singer [2008]), the response from such a tool, together with our empirical fit, can give engineers and geoscientists an approximation of the matrix permeabilities in the Rotokawa Andesite.

Permeability and Acoustic Velocities

There is a clear inverse relationship between our measurements of permeability and P-wave velocity (Figure 11F) such that the more permeable the sample, the slower the compressional wave velocity. These findings are consistent with the findings of many other authors (e.g., Vinciguerra et al. [2005]; Chaki et al. [2008]; Nara et al. [2011]; Heap et al. [2014]). The correlation of such properties is an excellent tool for understanding the micro- and mesoscopic fracture networks and their relation to permeability in the Rotokawa Andesite as follows: (1) we have shown that the porosity and crack density are closely linked (Figure 8A), (2) acoustic velocity and crack density are closely linked (Figure 8D), and (3) there is a power law correlation of Vp and permeability (Figure 11F). Thus, there is a direct link of P-wave velocity to permeability that is reliant on the crack densities of the samples. The relationship we present in Figure 11F shows a power-law fit which would indicate that the hydraulic radii of the pore space (pore and cracks) are similar in size but that the higher the concentration of cracks, the higher the permeability we observe (Bourbie and Zinszner [1985]).

Similarly, there are occasional mesofractures (with apertures less than 1-mm width; we note that these fractures are much smaller than those described in Massiot et al. [2012]) in the samples that deviate from the rest of the dataset (black stars, Figure 11F). The presence of these macrofractures increases permeability (by a factor of 2) and also appears deleterious to elastic wave propagation (all the three samples containing mesofractures have low elastic wave velocities, although we cannot separate the influence of meso- and microcracks on the velocities of these samples). Further, elastic waves are useful for the detection of cracks in rock and concrete (Chaki et al. [2008]; Heap et al. [2013]), and a decreased elastic wave velocity correlates well to more permeable media which is observed by the three outlying, higher permeability, lower elastic wave velocity samples.

The correlation between elastic wave velocity and permeability outside the laboratory has potentially far-reaching value for the prediction of reservoir permeability interactions from wireline logging and larger-scale seismic and microseismic surveys. There is a complex microseismic network installed at Rotokawa, and the location of earthquake activity has been closely linked to macroscopic permeability within the reservoir (Sewell et al. [2013]; Sherburn et al. [2013]). The existing model of the velocity structure at depth could then be further refined using our acoustic velocity and permeability data for reservoir rock matrix. This may allow a deeper and more accurate understanding of the distribution of permeability at depth.

Additionally, the data we have presented can also be used to infer values of matrix permeability from acoustic wireline logs (dipole sonic) used during exploration at nearby Ngatamariki Geothermal Field (Wallis et al. [2009]). Should similar geophysical logging be used in future wells drilled at Rotokawa, the matrix permeability may be estimated using the relationship we present here. In addition, the coupling of these data with microseismic data could allow a significant increase in understanding the complexity of the Rotokawa Andesite reservoir. While we are aware that macrofractures augment the elastic wave velocity during routine acoustic profiling (e.g., Barton and Zoback [1992]), our laboratory data show that although samples containing mesofractures (i.e., on the sample scale) are shifted to higher permeabilities and elastic wave velocities, they do not stray too far away from the trend extrapolated from our power-law relationship. Despite this, we urge a certain degree of caution, based on the potential presence of large-scale fractures, when estimating permeability using our derived permeability-elastic wave velocity relationship.

Application of Micromechanical and Geometrical Permeability Models

Extracting empirical relationships between laboratory-derived rock properties is useful; however, the parameters are not easily related to independently measurable quantities (i.e., they lack a physical basis). Micromechanical (e.g., the wing-crack model of Ashby and Sammis [1990]) and geometrical permeability models (e.g., the Kozeny-Carman relation, Guéguen and Palciauskas[1994]) can be better constrained as the parameters used in such models have a clear physical meaning. In this section, we attempt both sliding wing-crack modeling and Kozeny-Carman permeability modeling to investigate the microstructural controls on deformation and fluid flow, respectively.

Micromechanical Modeling

Micromechanical modeling can provide useful insights in the mechanics of compressive failure in brittle rock (Wong and Baud [2012]). Since the rocks of this study contain high microcrack densities, we will use the sliding wing-crack model of Ashby and Sammis ([1990]). This model idealizes the rock microstructure as an elastic continuum embedded with inclined (45°) microcracks (of length 2c). These microcracks act as stress concentrators for the initiation of 'wing' cracks when the frictional resistance of the closed crack is overcome and the stress at the tip of the crack exceeds the critical stress intensity factor (K_{IC}). The cracks can then propagate in the direction of the maximum principal stress. Eventually, the cracks coalesce, resulting in the failure of the elastic medium. In the case of uniaxial compression, Baud et al. ([2014]) derived an analytical approximation to estimate UCS:

$$\mathrm{UCS} = \frac{1.346}{\sqrt{1+\mu^2}-\mu} \; \frac{K_{\mathrm{Ic}}}{\sqrt{\pi c}} D_0^{-0.256} \tag{10}$$

Where μ is the friction coefficient of the sliding crack and D_0 is an initial damage parameter that is a function of the angle of the initial microcrack with respect to the maximum principal stress and the initial number of sliding cracks per unit area (Ashby and Sammis [1990]).

The analytical solution (that assumes an initial crack angle of 45°) presented above contains five parameters. We have, through experimental data and observations, a good handle on three of the parameters: (1) we have measured the UCS of 22 samples (Table 3), (2) μ rarely deviates from 0.6 to 0.7 (Byerlee [1978]), and (3) c can be determined from optical microscopy (we determined$c$ by measuring the approximate average length of the microcracks under the microscope). We do not have a laboratory-determined value for K_{IC}. While the K_{IC} of andesite has been previously measured to be about 1.5 $\mathrm{MPam}^{0.5}$ (Ouchterlony [1990]; Obara et al. [1992]; Tutluoglu and Keles[2011]; Nara et al. [2012]), there is no guarantee that this value is representative of the Rotokawa Andesite, which is likely to be lower than these values due to hydrothermal alteration. We therefore have chosen a slightly lower K_{IC} of 1.0 MPam$^{0.5}$ for our analysis. Using our UCS data, we can solve Equation 10 to assign a value of D_0 to each experiment (using $\mu = 0.6$; $K_{IC} = 1.0$; $c = 0.001$ m). The goal of such analysis, assuming that the other parameters remain roughly constant between different samples/cores, is to estimate D_0 using an easily measured physical property, such as Vp (therefore allowing us to predict rock strength, using the micromechanical model, from Vp measurements alone). Our analysis shows that D_0 ranges from 0.0019 to 0.26 for the 22 measured samples (with average of 0.039). D_0 is plotted against the crack area per unit volume (Sv) and Vp in Figure 12 and indicates that D_0 increases as Sv increases (Figure 12A). While this may appear logical (D_0 is a function of the initial crack density), it serves as an encouraging proof of the concept. The increase in D_0 with crack density is not linear; D_0 increases more rapidly beyond 10 mm^{-1} (Figure 12A). We also see that Vp decreases with increasing D_0; in detail, Vp decreases rapidly as D_0 increases from 0 to 0.05 and then decreases more gradually above 0.05. Unfortunately, the relationship between D_0 and Vp is a little more clouded (the data are more scattered, Figure 12B) and probably represents variable vesicle density (the model assumes that vesicles do not play a role in failure in compression) and hydrothermal alteration (we assume thatK_{IC} and the average crack lengths are constant). The conclusion of this pilot analysis is that the variability within the Rotokawa Andesite is potentially too large to permit meaningful microstructural wing-crack modeling, but greater success could be achieved with laboratory-determined values for K_{IC}. Therefore, if micromechanical modeling is to be deployed as a feasible method to predict the strength of Rotokawa Andesite reservoir rocks, the samples/cores should be grouped by their alteration, and K_{IC} measured for each alteration group.

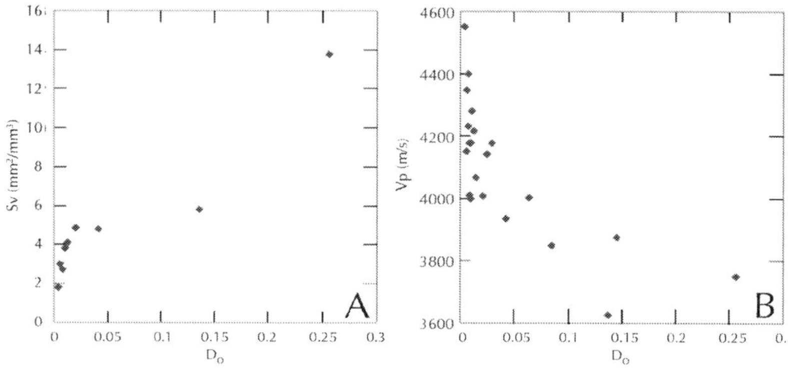

Figure 12: Results of geometric modeling for Rotokawa Andesite. (A) Initial damage parameter D_0 as predicted by Equation 10 and described by Baud et al. ([2014]) plotted versus calculated crack densities by the method of (Underwood [1970]). (B) Prediction of compressional wave velocity (Vp) as a function of the initial damage parameter D_0 the relationship between $D_0$and Vp shows a moderate correlation with high initial damage parameter but becomes quite clouded in those samples with a very small calculated D_0 (see text for further expansion on this relationship).

Permeability Modeling

Kozeny-Carman models are those that use the notion of a hydraulic radius (see Guéguen and Palciauskas [1994]) to correlate porosity and permeability. Forms of the Kozeny-Carman relation have previously been used in the study of volcanic rocks (e.g., Saar and Manga [1999]; Costa[2006]; Bernard et al. [2007]; Heap et al. [2014]), while others have used a heavily simplified version (e.g., Rust et al. [2003]; Mueller et al. [2005]; Lavallée et al. [2013]). The Kozeny-Carman relation is of the form:

$$k_{\mathrm{KC}} = \frac{\varphi(r_{\mathrm{H}})^2}{b\tau^2} \tag{11}$$

where k_{KC} is the permeability, φ is the connected porosity, b is a geometrical factor, τ is the tortuosity of the equivalent channel (i.e., the ratio of its actual to nominal length), and r_H is the hydraulic radius (i.e., the volume of pores divided by the surface of the pores). The power law exponent for our data (excluding those samples with macrofractures) is about 2.2 (Figure 11E) and is therefore consistent with the Kozeny-Carman model (Bourbie and Zinszner [1985]; Doyen[1988]). In detail, one would expect a power law exponent of 2 or 3 if the elements controlling the permeability are tubes or cracks, respectively (Guéguen and Palciauskas [1994]). Our power law exponent is between these two values. This is somewhat surprising, considering the pervasive fracture network in these materials, but could reflect flow through a combination of cracks and tubes or our limited porosity range. Since the entire dataset can be described by a single power law exponent, we conclude that within our limited range of connected porosities, there is no dramatic shift in pore space connectivity or tortuosity, as was the case for Fontainebleau sandstone at a porosity of 9 vol% (Bourbie and Zinszner [1985]) and andesite samples from Volcán de Colima (Mexico) at a porosity of about 11 vol% (Heap et al. [2014]). Extrapolating to porosities outside this range may be treacherous especially to lower porosities where samples may become subject to a higher power law exponent. However, within the dataset, the model predicts an increase in permeability of a factor of 1.5 for an increase in porosity of 1 vol% (an increase not uncommon for rock following a thermal stressing episode; e.g., Chaki et al. [2008]).

Application of Results to Geothermal Exploration and Utilization

The relationships between porosity, acoustic wave velocities, strength, and permeability are valuable for understanding a geothermal reservoir. Our data indicate strong correlations between these parameters, as observed by Stimac et al. ([2004], [2008]) amongst others. The data we have obtained are from cores sourced from three production wells. Such materials are very expensive to obtain, time consuming, and, if coring did not go as planned, can pose great risk of losing the well (Finger and Blankenship [2010]; Hole [2013]). The microstructural and empirical correlations presented in this study can be applied to new wells drilled in geothermal environments and can help refine studies on pre-existing wells, if our correlations hold true at the reservoir scale. Some physical parameters, such as porosity and elastic wave velocities, are easily obtainable through the use of down-hole geophysical logging suites. The empirical correlations shown in this study (bolstered by our application of classical models) show that readily measurable physical properties may therefore be used to predict more complex and pertinent properties such as strength and permeability. Such correlations and calibrations are common in the hydrocarbon industry especially during exploration drilling (e.g., Vernik et al. [1993] and references therein), and we consider that our dataset can help improve the understanding of the Rotokawa reservoir while minimizing the risk to future drilling operations.

A clear understanding of the factors that control reservoir rock permeability is fundamental for the planning of stimulation and enhancement operations that may be necessary as the Rotokawa field and reservoir dynamics change with continued production. The need to drill additional wells or re-work pre-existing wells may become apparent and the ease at which the reservoir can accept and deliver fluids (i.e., its permeability) will be of utmost importance. The thermal stimulation of injection wells has taken place at Rotokawa for some time by the injection of power-plant condensates and spent brines (Siega et al. [2009]), but the technique may play a significant role in enhancing production wells at some future stage.

Therefore, a deeper understanding of how permeability may be increased through stimulation is important. The application of models such as the Kozeny-Carman may provide insight to permeability enhancement. An increase in the porosity of reservoir rock by 1 vol%, according to the geometrical model, should increase the permeability by a factor of 1.5. In the case of an aging field and aging wellbores, such an increase could greatly extend the life of the field. In the interests of keeping geothermal projects commercially economic, the fundamental understanding of the reservoir rock properties become essential to the continued utilization and management of the field.

CONCLUSIONS

Our study provides a comprehensive evaluation of the physical and mechanical properties of the Rotokawa Andesite through a multi-disciplinary approach. We have evaluated the Rotokawa Andesite from the microstructural to macroscopic scale and have presented robust datasets that permit the correlation and comparison of important physical properties to geothermal exploitation. A comprehensive understanding of how the relationships of microstructural texture influence key physical properties such as strength and permeability, essential for the optimal utilization of a geothermal resource have been investigated.

Further, we summarize our conclusions as follows:

- We have shown that the presence and intensity of microfracturing in the Rotokawa Andesite are the predominant controlling factors on physical and mechanical properties. The behavior of these properties is also shown to be largely independent of the alteration mineralogy as we see similar alteration intensities in the samples we have studied.
- Guided by a systematic understanding of role of microfractures, we show that empirical correlations of strength and porosity can be developed and applied to field scale engineering problems. We have shown that as the porosity increases, the strength decreases and elastic wave velocities are attenuated. Similarly, we show that permeability increases with increased porosity and reduced acoustic velocity. These findings are applicable if geophysical logging tools be used after the drilling of wells to ascertain properties such as porosity; our dataset provides useful means to address complex reservoir problems.
- We further boost our empirical correlations by applying classical physical models based on sound physical theory to predict both UCS and permeability through understanding of the microstructure. We have applied these models with some success, but these models are best-suited for homogeneous, isotropic materials. Further work to constrain these models should include laboratory investigations of fracture toughness (K_{IC}) and the factors that influence this variable. However, our fit for the damage criterion D_0 is acceptable and builds the foundations for future understanding and may permit the construction of similar better constrained models.
- The study comprises a large dataset with a goal to further push the knowledge that can be sourced from a geothermal environment such as the Rotokawa Andesite. The properties that we have evaluated are very difficult to constrain without direct information from rocks sourced from the reservoir. Geothermal reservoirs

are complex, and harsh environments from which the recovery of intact core can present a significant and financially risky challenge. The results that we present here help us to understand this complex reservoir environment by their application to field scale engineering and geological issues.

- Our analyses have provided quantifiable and measurable physical properties of the Rotokawa Andesite. However, the dataset is not exhaustive. Further studies need to be carried out to replicate near-reservoir conditions in the laboratory and should focus on permeability at the high confining pressures and temperatures found in the reservoir. Additionally, mechanical testing such as triaxial, tensile strength, and fracture toughness experiments should be conducted under high-temperature conditions, potentially in the presence of reservoir-type fluids to aid in predictions of reservoir behavior and geomechanical modeling under conditions as close as possible to those found in the reservoir.

AUTHORS' CONTRIBUTIONS

PS performed the majority of the experimental work and composition of the manuscript. MH assisted in revision and suggestions for the manuscript and performed the permeability and micromechanical modeling. MV provided assistance in experimental work and processing of mechanical data. JC provided revision and petrological input. TR provided the permeability measurements. All authors read and approved the final manuscript.

ACKNOWLEDGEMENTS

The authors wish to thank Mighty River Power Company Ltd. for a generous grant for PS, which allowed collaboration with MH and TR. We also wish to thank the Rotokawa Joint Venture, a joint venture between the Tauhara North No. 2 Trust and Mighty River Power Company Ltd. for the core material used in this study. The staff of the Department of Geological Sciences at the University of Canterbury were invaluable in assisting in all aspects of this research. The Brian

Mason Trust also provided for the transportation and delivery of the core to UC. The authors of this study also acknowledge a Hubert Curien Partnership (PHC) Dumont d'Urville grant (grant number 31950RK) which has assisted the France-New Zealand collaboration for this and future projects. MH was partly funded by the LABEX ANR-11-LABX-0050_G-EAU-THERMIE-PROFONDE framework (funding from the state managed by the French National Research Agency as part of the Investments for the Future Program).

REFERENCES

1. Al-Harthi AA, Al-Amri RM, Shehata WM (1999) The porosity and engineering properties of vesicular basalt in Saudi Arabia. Eng Geol 54:313-320
2. Ashby MF, Sammis CG (1990) The damage mechanics of brittle solids in compression. Pure Appl Geophys 133:489-521
3. Barton CA, Zoback MD (1992) Self-similar distribution and properties of macroscopic fractures at depth in crystalline rock in the Cajon Pass Scientific Drill Hole. J Geophys Res 97(B4):5181-5200
4. Baud P, Wong T-F, Zhu W (2014) Effects of porosity and crack density on the compressive strength of rocks. Int J Rock Mech Min Sci : (in press)
5. Bernard M-L, Zamora M, Géraud Y, Boudon G (2007) Transport properties of pyroclastic rocks from Montagne Pelée volcano (Martinique, Lesser Antilles). J Geophys Res 112:
6. Bibby HM, Caldwell TG, Davey F, Webb T (1995) Geophysical evidence on the structure of the Taupo Volcanic Zone and its hydrothermal circulation. J Volcanol Geotherm Res 68:29-58
7. Bieniawski ZT (1967) Mechanism of brittle fracture of rock part II - experimental studies. Int J Rock Mech Min Sci 4:407-423
8. Blake OO, Faulkner DR, Rietbrock A (2012) The effect of varying damage history in crystalline rocks on the P- and S-wave velocity under hydrostatic confining pressure. Pure Appl Geophys 170:493-505
9. Bloomberg S, Rissmann C, Mazot A, Oze C, Horton T, Kennedy B, Werner C, Christenson B, Pawson J (2012) Soil gas flux

exploration at the Rotokawa Geothermal Field and White Island, New Zealand. In: Proceedings, Thirty Sixth Workshop on Geothermal Reservoir Engineering. Stanford University, Stanford, California.30 January 30 to 1 February 2012

10. Bourbie T, Zinszner B (1985) Hydraulic and acoustic properties as a function of porosity in Fontainebleau sandstone. J Geophys Res 90:11524-11532
11. Brace WF, Bombolakis EG (1963) A note on brittle crack growth in compression. J Geophys Res 68:3709-3713
12. Brace WF, Paulding B, Scholz C (1966) Dilatancy in the fracture of crystalline rocks. J Geophys Res 71:3939-3953
13. Brace WF, Walsh JB, Frangos WT (1968) Permeability of granite under high pressure. J Geophys Res 73:2225-2236
14. Byerlee JD (1978) Friction of rocks. Pure Appl Geophys 116:615-626
15. Chaki S, Takarli M, Agbodjan WP (2008) Influence of thermal damage on physical properties of a granite rock: porosity, permeability and ultrasonic wave evolutions. Constr Build Mater 22:1456-1461
16. Chang C, Zoback MD, Khaksar A (2006) Empirical relations between rock strength and physical properties in sedimentary rocks. J Pet Sci Eng 51:223-237
17. Cole JW (1990) Structural control and origin of volcanism in the Taupo volcanic zone, New Zealand. Bull Volcanol 52:445-459
18. Collar RJ, Browne PRL (1985) Hydrothermal eruptions at The Rotokawa Geothermal Field, Taupo Volcanic Zone, New Zealand. In: Proceedings of the seventh New Zealand geothermal workshop, University of Auckland. Geothermal Institute, Auckland, New Zealand 6–8 November 1985
19. Costa A (2006) Permeability-porosity relationship: a re-examination of the Kozeny-Carman equation based on a fractal pore-space geometry assumption. Geophys Res Lett 33:
20. David C, Menendez B, Darot M (1999) Influence of stress-induced and thermal cracking on physical properties and microstructure of La Peyratte granite. Int J Rock Mech Min Sci 36:433-448
21. Diamantis K, Gartzos E, Migiros G (2009) Study on uniaxial compressive strength, point load strength index, dynamic and

physical properties of serpentinites from Central Greece: test results and empirical relations. Eng Geol 108:199-207

22. Diederichs M, Kaiser P, Eberhardt E (2004) Damage initiation and propagation in hard rock during tunneling and the influence of near-face stress rotation. Int J Rock Mech Min Sci 41:785-812
23. DiPippo R (2008) Geothermal power plants: principles, applications, case studies and environmental impact. Elsevier Ltd, Oxford.
24. Doyen PM (1988) Permeability, conductivity, and pore geometry of sandstone. J Geophys Res 93:7729-7740
25. Eberhardt E, Stead D, Stimpson B, Read RS (1998) Identifying crack initiation and propagation thresholds in brittle rock. Can Geotech J 35:222-233
26. Ellis DV, Singer JM (2008) Well logging for earth scientists. Springer, Dordrecht.
27. Faoro I, Vinciguerra S, Marone C, Elsworth D, Schubnel A (2013) Linking permeability to crack density evolution in thermally stressed rocks under cyclic loading. Geophys Res Lett 40:2590-2595
28. Ferrero AM, Marini P (2001) Experimental studies on the mechanical behaviour of two thermal cracked marbles. Rock Mech Rock Eng 34:57-66
29. Finger J, Blankenship D (2010) Handbook of best practices for geothermal drilling. Sandia National Laboratories, Albuquerque.
30. Fredrich JT, Wong T (1986) Micromechanics of thermally induced cracking in three crustal rocks. J Geophys Res 91:12743-12764
31. Géraud Y (1994) Variations of connected porosity and inferred permeability in a thermally cracked granite. Geophys Res Lett 21:979-982
32. Grant MA, Bixley PF (2011) Geothermal reservoir engineering. Elsevier Science Ltd, Oxford.
33. Guéguen Y, Palciauskas V (1994) Introduction to the physics of rocks. Princeton University Press, Princeton.
34. Guéguen Y, Schubnel A (2003) Elastic wave velocities and permeability of cracked rocks. Tectonophysics 370:163-176
35. Gupta H, Sukanta R (2006) Geothermal energy: an alternative resource for the 21st century. Elsevier B.V., Oxford.

36. Hardy H (1981) Applications of acoustic emission techniques to rock and rock structures: a state of the art review. In: Drnevich V, Gray R (eds) Acoustic emission in geotechnical engineering practice, American Society for Testing and Materials, University of Michigan, Ann Arbor.
37. Heap MJ, Faulkner DR (2008) Quantifying the evolution of static elastic properties as crystalline rock approaches failure. Int J Rock Mech Min Sci 45:564-573
38. Heap MJ, Vinciguerra S, Meredith PG (2009) The evolution of elastic moduli with increasing crack damage during cyclic stressing of a basalt from Mt. Etna volcano. Tectonophysics 471:153-160
39. Heap MJ, Lavallée Y, Laumann A, Hess K-U, Meredith PG, Dingwell DB, Huismann S, Weise F (2013) The influence of thermal-stressing (up to 1000°C) on the physical, mechanical, and chemical properties of siliceous-aggregate, high-strength concrete. Construct Build Mater 42:248-265
40. Heap MJ, Lavallee Y, Petrakova L, Baud P, Reushcle T, Varley NR, Dingwell DB (2014) Microstructural controls on the physical and mechanical properties of edifice-forming andesites at Volcán de Colima Mexico. J Geophys Res 119:2925-2963
41. Heard HC, Page L (1982) Elastic moduli, thermal expansion, and inferred permeability of two granites to 350°C and 55 megapascals. J Geophys Res 87:9340-9348
42. Hedenquist JW, Mroczek EK, Giggenbach WF (1988) Geochemistry of the Rotokawa geothermal system: summary of data, interpretation and appraisal for energy development. In: Chemistry Division DSIR Technical Note 88/6. , .
43. Hole HM (2013) Geothermal drilling - keep it simple. In: Proceedings of the 35th New Zealand geothermal workshop. , Rotorua, New Zealand.17–20 November 2013
44. Horie T, Muto T (2010) The world's largest single cylinder geothermal power generation unit - Nga Awa Purua Geothermal Power Station, New Zealand. Geothermal Res Council Trans 34:1039-1044
45. Jafari A, Babadagli T (2011) Effective fracture network permeability of geothermal reservoirs. Geothermics 40:25-38

46. Jaya MS, Shapiro SA, Kristinsdóttir LH, Bruhn D, Milsch H, Spangenberg E (2010) Temperature dependence of seismic properties in geothermal rocks at reservoir conditions. Geothermics 39:115-123
47. Ju Y, Yang Y, Peng R, Mao L (2013) Effects of pore structures on static mechanical properties of sandstone. J Geotech Geoenvironmental Eng 139:1745-1755
48. Kahraman S (2001) Evaluation of simple methods for assessing the uniaxial compressive strength of rock. Int J Rock Mech Min Sci 38:981-994
49. Kahraman S, Gunaydin O, Fener M (2005) The effect of porosity on the relation between uniaxial compressive strength and point load index. Int J Rock Mech Min Sci 42:584-589
50. Keshavarz M, Pellet FL, Loret B (2010) Damage and changes in mechanical properties of a gabbro thermally loaded up to 1,000°C. Pure Appl Geophys 167:1511-1523
51. Klinkenberg LJ (1941) The permeability of porous media to liquids and gases. Drilling and production practice. American Petroleum Institute, New York.
52. Kristinsdóttir LH, Flóvenz ÓG, Árnason K, Bruhn D, Milsch H, Spangenberg E, Kulenkampff J (2010) Electrical conductivity and P-wave velocity in rock samples from high-temperature Icelandic geothermal fields. Geothermics 39:94-105
53. Krupp RE, Seward TM (1987) The Rotokawa geothermal system, New Zealand; an active epithermal gold-depositing environment. Econ Geol 82:1109-1129
54. Lavallée Y, Benson PM, Heap MJ, Hess KU, Flaws A, Schillinger B, Meredith PG, Dingwell DB (2013) Reconstructing magma failure and the degassing network of dome-building eruptions. Geology 41:515-518
55. Legmann H, Sullivan P (2003) The 30 MW Rotokawa I geothermal project five years of operation. In: International geothermal conference. , Reykjavik, Iceland September 2003
56. Li L, Aubertin M (2003) A general relationship between porosity and uniaxial strength of engineering materials. Can J Civ Eng 30:644-658

57. Lim SS, Martin CD, Åkesson U (2012) In-situ stress and microcracking in granite cores with depth. Eng Geol 147–148:1-13

58. Lion M, Skoczylas F, Ledésert B (2005) Effects of heating on the hydraulic and poroelastic properties of Bourgogne limestone. Int J Rock Mech Min Sci 42:508-520

59. Luping T (1986) A study of the quantitative relationship between strength and pore-size distribution of porous materials. Cem Concr Res 16:87-96

60. Lutz SJ, Hickman S, Davatzes N, Zemach E, Drakos P, Robertson-Tait A (2010) Rock mechanical testing and petrologic analysis in support of well stimulation activities at the Desert Peak Geothermal Field, Nevada. In: Proceedings of the thirty-fifth workshop on geothermal reservoir engineering. Stanford University, Stanford, California. 1–3 February 2010

61. Martin CD (1993) The strength of Massive Lac du Bonnet granite around underground openings. Dissertation, University of Manitoba.

62. Martin CD, Chandler NA (1994) The progressive fracture of Lac du Bonnet granite. Int J Rock Mech Min Sci 31:643-659

63. Martínez-Martínez J, Benavente D, García-del-Cura MA (2011) Spatial attenuation: the most sensitive ultrasonic parameter for detecting petrographic features and decay processes in carbonate rocks. Eng Geol 119:84-95

64. Massiot C, McNamara D, Lewis B, Price L, Bignall G (2012) Statistical corrections of fracture sampling bias in boreholes from acoustic televiewer logs. In: New Zealand geothermal workshop proceedings. , Auckland, New Zealand. 19–21 November 2012

65. Mueller S, Melnik O, Spieler O (2005) Permeability and degassing of dome lavas undergoing rapid decompression: an experimental determination. Bull Volcanol 67:526-538

66. Nara Y, Meredith PG, Yoneda T, Kaneko K (2011) Influence of macro-fractures and micro-fractures on permeability and elastic wave velocities in basalt at elevated pressure. Tectonophysics 503:52-59

67. Nara Y, Morimoto K, Hiroyoshi N, Yoneda T, Kaneko K, Benson PM (2012) Influence of relative humidity on fracture toughness of

rock: implications for subcritical crack growth. Int J Solids Struct 49:2471-2481

68. Nicksiar M, Martin CD (2012) Evaluation of methods for determining crack initiation in compression tests on low-porosity rocks. Rock Mech Rock Eng 45:607-617
69. Obara Y, Sakaguchi K, Nakayama T, Sugawara K (1992) Anisotropy effect on fracture toughness of rocks. Int J Rock Mech Min Sci Geomech 30:137
70. Ouchterlony F (1990) Fracture toughness testing of rock with core based specimens. Eng Fract Mech 35:351-366
71. Palchik V (2013) Is there a link between the type of the volumetric strain curve and elastic constants, porosity, stress and strain characteristics? Rock Mech Rock Eng 46:315-326
72. Palchik V, Hatzor YH (2002) Crack damage stress as a composite function of porosity and elastic matrix stiffness in dolomites and limestones. Eng Geol 63:233-245
73. Pereira J-M, Arson C (2013) Retention and permeability properties of damaged porous rocks. Comput Geotech 48:272-282
74. Pola A, Crosta G, Fusi N, Barberini V, Norini G (2012) Influence of alteration on physical properties of volcanic rocks. Tectonophysics 566–567:67-86
75. Pola A, Crosta GB, Fusi N, Castellanza R (2014) General characterization of the mechanical behaviour of different volcanic rocks with respect to alteration. Eng Geol 169:1-13
76. Powell T (2011) Natural subsidence at the Rotokawa Geothermal Field and implications for permeability development. Geothermal Res Council Trans 35:973-976
77. Quinao J, Sirad-Azwar L, Clearwater J, Hoepfinger V, Le Brun M, Bardsley C (2013) Analyses and modeling of reservoir pressure changes to interpret the Rotokawa Geothermal Field response to Nga Awa Purua Power Station operation. In: Proceedings of the 38th workshop on geothermal reservoir engineering. Stanford University, Stanford, California. 11–13 February 2013
78. Rae A (2007) Rotokawa geology and geophysics. GNS Science consultancy report 2007/83 May 2007. GNS Science, Lower Hutt.

79. Rae AJ, McCoy-West AJ, Ramirez LE, Alcaraz SA (2009) Geology of production well RK28. Rotokawa Geothermal Field. GNS Science consultancy report 2009/253 September 2009. GNS Science, Lower Hutt.
80. Rae AJ, McCoy-West AJ, Ramirez LE, McNamara D (2010) Geology of production wells RK30L1 and RK30L2, Rotokawa Geothermal Field. GNS Science consultancy report 2010/02 January 2010. GNS Science, Lower Hutt.
81. Rajabzadeh MA, Moosavinasab Z, Rakhshandehroo G (2011) Effects of rock classes and porosity on the relation between uniaxial compressive strength and some rock properties for carbonate rocks. Rock Mech Rock Eng 45:113-122
82. Ramirez LE, Hitchcock D (2010) Geology of production well RK27L2, Rotokawa Geothermal Field, GNS Science Consultancy Report 2010/100 April 2010. GNS Science, Lower Hutt.
83. Reuschlé T, Gbaguidi Haore S, Darot M (2006) The effect of heating on the microstructural evolution of La Peyratte granite deduced from acoustic velocity measurements. Earth Planet Sci Lett 243:692-700
84. Richter D, Simmons G (1977) Microcracks in crustal igneous rocks: microscopy. In: Heacock JG, Keller GV, Oliver JE, Simmons G (eds) The earth's crust, American Geophysical Union, Washington, DC.
85. Rowland JV, Sibson RH (2004) Structural controls on hydrothermal flow in a segmented rift system, Taupo Volcanic Zone, New Zealand. Geofluids 4:259-283
86. Rowland JV, Wilson CJN, Gravley DM (2010) Spatial and temporal variations in magma-assisted rifting, Taupo Volcanic Zone, New Zealand. J Volcanol Geotherm Res 190:89-108
87. Rust AC, Manga M, Cashman KV (2003) Determining flow type, shear rate and shear stress in magmas from bubble shapes and orientations. J Volcanol Geotherm Res 122:111-132
88. Rutter EH (1986) On the nomenclature of mode of failure transitions in rocks. Tectonophysics 122:381-387
89. Saar MO, Manga M (1999) Permeability-porosity relationship in vesicular basalts. Geophys Res Lett 26:111-114

90. Schöpfer MPJ, Abe S, Childs C, Walsh JJ (2009) The impact of porosity and crack density on the elasticity, strength and friction of cohesive granular materials: insights from DEM modelling. Int J Rock Mech Min Sci 46:250-261
91. Sewell SM, Cumming WB, Azwar L, Bardsley C (2012) Integrated MT and natural state temperature interpretation for a conceptual model supporting reservoir numerical modelling and well targeting at the Rotokawa Geothermal Field, New Zealand. In: Proceedings of the thirty-seventh workshop on geothermal reservoir engineering. Stanford University, Stanford California. 30 January to 1 February 2012
92. Sewell SM, Cumming W, Bardsley CJ, Winick J, Quinao J, Wallis IC, Sherburn S, Bourguignon S, Bannister S (2013) Interpretation of microearthquakes at the Rotokawa Geothermal Field, 2008 to 2012. In: Proceedings of the 35th New Zealand geothermal workshop. , Rotorua, New Zealand. 17–20 November 2013
93. Sherburn S, Bourguignon S, Bannister S, Sewell S, Cumming B, Bardsley C, Quinao J, Wallis I (2013) Microseismicity at Rotokawa Geothermal Field, 2008 to 2012. In: Proceedings of the 35th New Zealand geothermal workshop. , Rotorua, New Zealand. 17–20 November 2013
94. Siega C, Grant M, Powell T (2009) Enhancing injection well performance by cold water stimulation in Rotokawa and Kawerau geothermal field. In: Proceedings of PNOC-EDC conference. , Manila, Philippines. 27–28 September 2009
95. Smith R, Sammonds PR, Kilburn CRJ (2009) Fracturing of volcanic systems: experimental insights into pre-eruptive conditions. Earth Planet Sci Lett 280:211-219
96. Sousa LMO, Suárez del Río LM, Calleja L, Ruiz de Argandoña VG, Rey AR (2005) Influence of microfractures and porosity on the physico-mechanical properties and weathering of ornamental granites. Eng Geol 77:153-168
97. Stanchits S, Vinciguerra S, Dresen G (2006) Ultrasonic velocities, acoustic emission characteristics and crack damage of basalt and granite. Pure Appl Geophys 163:975-994
98. Stimac JA, Powell TS, Golla GU (2004) Porosity and permeability of the Tiwi geothermal field, Philippines, based on continuous and spot core measurements. Geothermics 33:87-107

99. Stimac J, Nordquist G, Suminar A, Sirad-Azwar L (2008) An overview of the Awibengkok geothermal system, Indonesia. Geothermics 37:300-331
100. Takarli M, Prince W, Siddique R (2008) Damage in granite under heating/cooling cycles and water freeze thaw condition. Int J Rock Mech Min Sci 45:1164-1175
101. Tugrul A, Gurpinar O (1997) the effect of chemical weathering on the engineering properties of Eocene basalts in northeastern Turkey. Environ Eng Geosci 3:225-234
102. Tutluoglu L, Keles C (2011) Mode I fracture toughness determination with straight notched disk bending method. Int J Rock Mech Min Sci 48:1248-1261
103. Ulusay R, Hudson J (2007) the complete ISRM suggested methods for rock characterization, testing and monitoring: 1974–2006. Elsevier, Antalya, Turkey.
104. Underwood EE (1970) Quantitative stereology for microstructural analysis. In: Underwood EE (ed) Quantitative stereology, Addison-Wesley, Reading, Massachusetts.
105. Vernik L, Bruno M, Bovberg C (1993) Empirical relations between compressive strength and porosity of siliciclastic rocks. Int J Rock Mech Min Sci Geomech 30:677-680
106. Vinciguerra S, Trovato C, Meredith P, Benson P (2005) Relating seismic velocities, thermal cracking and permeability in Mt. Etna and Iceland basalts. Int J Rock Mech Min Sci 42:900-910
107. Wallis I, McCormick S, Sewell S, Boseley C (2009) Formation assessment in geothermal using wireline tools - application and early results from the Ngatamariki Geothermal Field, New Zealand. In: Proceedings of the New Zealand Geothermal Workshop. , Rotorua, New Zealand. 16–18 November 2009
108. Walsh JB (1965) the effect of cracks on the compressibility of rock. J Geophys Res 70:381-389
109. Walsh JB (1965) the effect of cracks in rocks on Poisson's ratio. J Geophys Res 70:5249-5257
110. Wang HF, Bonner BP, Carlson SR, Kowallis BJ, Heard HC (1989) Thermal stress cracking in granite. J Geophys Res 94:1745-1758
111. Watanabe T, Shimizu Y, Noguchi S, Nakada S (2008) Permeability measurements on rock samples from Unzen Scientific Drilling

Project Drill Hole 4 (USDP-4). J Volcanol Geotherm Res 175:82-90

112. Wilson CJN, Houghton BF, Mcwilliams MO, Lanphere MA, Weaver SD, Briggs RM (1995) Volcanic and structural evolution of Taupo Volcanic Zone, New Zealand: a review. J Volcanol Geotherm Res 68:1-28
113. Wong TF, Baud P (2012) The brittle-ductile transition in porous rock: a review. J Struct Geol 44:25-53
114. Wu XY, Baud P, Wong TF (2000) Micromechanics of compressive failure and spatial evolution of anisotropic damage in Darley Dale sandstone. Int J Rock Mech Min Sci 37:143-160

Chapter 7

Seismic Tomography of the Area of the 2010 Beni-Ilmane Earthquake Sequence, North-Central Algeria

Issam Abacha[1], Ivan Koulakov[2, 3], Fethi Semmane[1], and Abd Karim Yelles-Chaouche[1]

[1]Centre de Recherche en Astronomie, Astrophysique et Geophysique, BP. 63, Bouzareah, Alger, Algeria

[2]Trofimuk Institute of Petroleum Geology and Geophysics, SB RAS, Prospekt Koptyuga, 3, Novosibirsk 630090, Russia

[3]Novosibirsk State University, Novosibirsk, Russia

ABSTRACT

The region of Beni-Ilmane (District of M'sila, north-central Algeria) was the site of an earthquake sequence that started on 14 May 2010. This sequence, which lasted several months, was triggered by conjugate E–W reverse and N–S dextral faulting. To image the crustal structure of these active faults, we used a set of 1406 well located aftershocks events and applied the local tomography software (LOTOS) algorithm, which includes absolute source location, optimization of the initial 1D velocity model, and iterative tomographic inversion for 3D seismic P- and S-wave velocities (and the Vp/Vs ratio), and source parameters. The patterns of P-wave low-velocity anomalies correspond to the alignments of faults determined from geological evidence, and the P-wave high-velocity anomalies may represent rigid blocks of the upper crust that are not deformed by regional stresses. The S-wave low-velocity anomalies coincide with the aftershock area, where relatively high values of Vp/Vs ratio (1.78) are observed compared with values in the surrounding areas (1.62–1.66). These high values may indicate high fluid contents in the aftershock area. These fluids could have been released from deeper levels by fault movements during earthquakes and migrated rapidly upwards. This hypothesis is supported by vertical sections across the study area show that the major Vp/Vs anomalies are located above the seismicity clusters.

INTRODUCTION

Northern Algeria has been the location of many destructive earthquakes, including the recent 21 May 2003 strong (M_w6.9) event at Boumerdes (e.g., Yelles et al. 2004). Several recent moderate seismic events have also occurred in this region, such as the 2006 Lâalam earthquake (M_w5.2) (Beldjoudi et al. 2009) in the Kherrata fault system north of Setif and the 2006 Tadjena earthquake (M_w5.0) (Beldjoudi et al. 2011). In addition, in 2010, an earthquake sequence occurred in the region of Beni-Ilmane (Yelles et al. 2013), a small village located in the southern part of the Tellian Atlas and about 200 km southeast of Algiers (Figure 1a). This earthquake sequence caused 3 fatalities and injured 170 people, and generated substantial damage to houses and local infrastructure in the epicentral area.

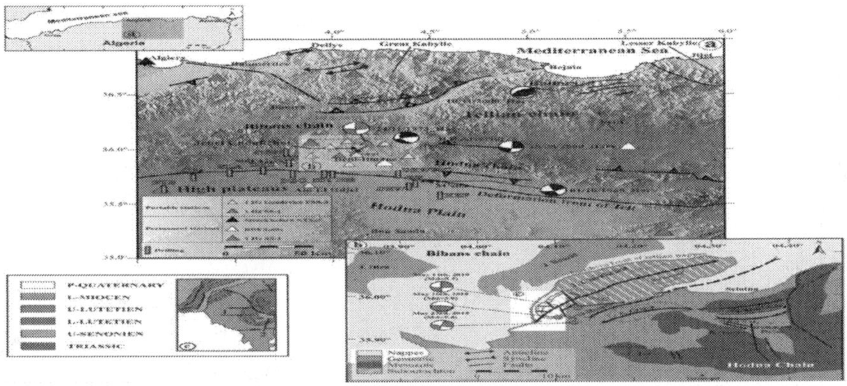

Figure 1: Seismo-tectonic context of the 2010 Beni-Ilmane seismic sequence (from Yelles et al.2013) with the seismic networks. The brown rectangles represent the drillings in the region and some among them attained around 4000 m of depth. The red dashed box shows the tomography area shown in Figures 3, 4 and 6.

An earthquake sequence such as the 2010 Beni-Ilmane sequence is uncommon in Algeria in several respects. First, the sequence consisted of three main shocks of very similar magnitudes ($5.0 < M_d < 5.2$) over a period of nine days in a localized area. Second, the sequence was characterized by strong aftershock activity (more than 24,000 events were recorded by a temporary seismic network during the first week of the sequence) that lasted for several months. Third, the earthquake sequence occurred in a region of lower seismic activity in northern Algeria, and more precisely between two geological transition zones; namely, the Bibans-Hodna Mountains and the Tellian Chain High Plateaus (Figure 1a).

Prior to the Beni-Ilmane sequence, the study region was characterized by low seismic activity, although some historical events have occurred here. The most notable event occurred close to the epicenter of the 2010 Beni-Ilmane event on 21 February 1960, and was of magnitude M5.6. This event generated similar macroseismic effects to those of the Beni-Ilmane event, and its maximum intensity was VIII (MSK) (Benouar 1994).

To understand the crustal structure in the Beni-Ilman region, and to enhance the accuracy of source locations there, we performed a tomographic inversion that uses the simultaneous determination of P-

and S-wave velocity distributions as well as the relocation of aftershocks. Local earthquake tomography (LET) is a useful tool for imaging lateral heterogeneities in the upper crust based on the relationship between the pattern of P- and S-wave velocity anomalies and the clustering of aftershocks along active fault zones. LET has been used for studying the distribution of seismic velocity in several major fault systems, including the San Andreas Fault in the western US (e.g., Chiarabba and Amato 1994; Dorbath et al. 1996; Thurber et al. 1997) and the North Anatolian Fault in Turkey (e.g., Salah et al. 2007, 2011; Koulakov et al. 2010; Yolsal-Çevikbilen et al. 2012).

Tomography studies in Algeria have been implemented mainly for studying aftershocks at three sites where the greatest number of large earthquakes have been recorded, namely in the region of the 1980 El-Asnam earthquake (Chiarabba et al. 1997), the region of the 1985 Constantine earthquake (Bounif et al. 1998), and the epicentral area of the 2003 Boumerdes earthquake (Ayadi et al. 2008). The seismic tomography of the El-Asnam region (west-central Algeria) shows a complex preexisting structure along which the evolution of a rupture is controlled by the heterogeneity of the crust (Chiarabba et al. 1997). Seismic tomography of the Constantine earthquake source region (eastern Algeria) has yielded the 3D P-wave velocity to depths of 12 km. The resulting seismic model revealed the locations of possible fault planes and a segmentation of the aftershocks into three ruptured segments (Bounif et al. 1998). For the Boumerdes earthquake region (offshore Algeria), the high-resolution study of aftershocks and velocity structure helped to constrain the fault geometry and structure (Ayadi et al. 2008).

SEISMOTECTONIC CONTEXT

In northern Algeria, seismic activity resulting from the convergence of the African and Eurasian plates is concentrated mainly in the Tellian chain, where moderate to large seismic events have occurred. The most seismically active region is the coastal part of the chain along the junction with the offshore part. Many large events have been recorded in this region, including the most recent earthquake in Boumerdes in 2003, and those documented in historical catalogues, such as the Oran event of 1790, the Algiers events of 1365 and 1716, and the Djidjelli event of 1856.

In the western and central parts of the country, seismicity is related to NE–SW reverse faults or NW–SE conjugate faults, which may be explained by regional compression due to the convergence of Africa with Eurasia. In the eastern part of the country, strike-slip faults are more common, as are found, for example, in the Mila and Guelma basins.

The Beni-Ilmane sequence took place in a transition zone between the Bibans and Hodna massifs, the two main geologic edifices of the Atlasic chain of northern Algeria. More precisely, the event occurred at the western tip of the Jebel Choukchot, one of the southernmost massifs of the Bibans chain.The Choukchot anticline is cross-cut by several E–W- and N–S-trending faults (Figure 1c). Towards the South, the syncline of Beni-Ilmane faces the Jebel Choukchot. It is marked in its NE extension by the existence of a nappes unit called the Selatna tongue (Figure 1b).

As part of the Saharan platform of Algeria, the South Tellian border region, including the Hodna-Bibans region (near Msila) has been of interest to the Algerian Oil Company for several decades. Oil exploration started in this part of Algeria more precisely in the Sidi Aissa region, about 40 km from Beni-Ilmane village, in 1950. Exploration were conducted in two periods: from 1950 to 1962 by Sn Repal (Société Nationale de Recherche et d'Exploitation des Pétroles en Algérie) and from 1963 to the present by Sonatrach (Société Nationale du Transport, de Recherche et de Commercialisation des Hydrocarbures). Seismic profiling and drilling (Figure 1a) performed during these two periods have revealed high hydrocarbon potential in the South Tellian border region. Investigations indicate that oil is contained in the Eocene strata and two reservoir levels have been distinguished. The "A" level is 10–15 m thick and corresponds to the upper Lutetian, and the "B" level (main reservoir) is 40–70 m thick and corresponds to the Thanetian–Ypresian (Kieken 1974 and 1975).

DATA

The data for the study were recorded by a temporary seismic network consisting of 11 three-component short-period portable stations deployed for about five months in the epicentral area of the 2010 Beni-Ilmane sequence, in addition to the permanent stations of the

Algerian seismic network (Figure 1b). Three stations of the permanent network are located at distances of less than 50 km from the epicentral area. For a description of the type and technical characteristics of these stations, see Yelles et al. (2013). The arrival times of P- and S waves from local events were manually picked using the Antelope software of Kinemetrics. Figure 2 shows an example of a seismic record of several aftershocks from three portable stations with indications of the P- and S-wave arrivals. To select the data, we used events with no less than 8 picks having the residuals of smaller than 0.8 s. A total of 1406 events were initially selected for the study, corresponding to 12,380 P-wave and 2604 S-wave picks for the interval between 14 and 31 May 2010 (the period of maximum aftershock activity).

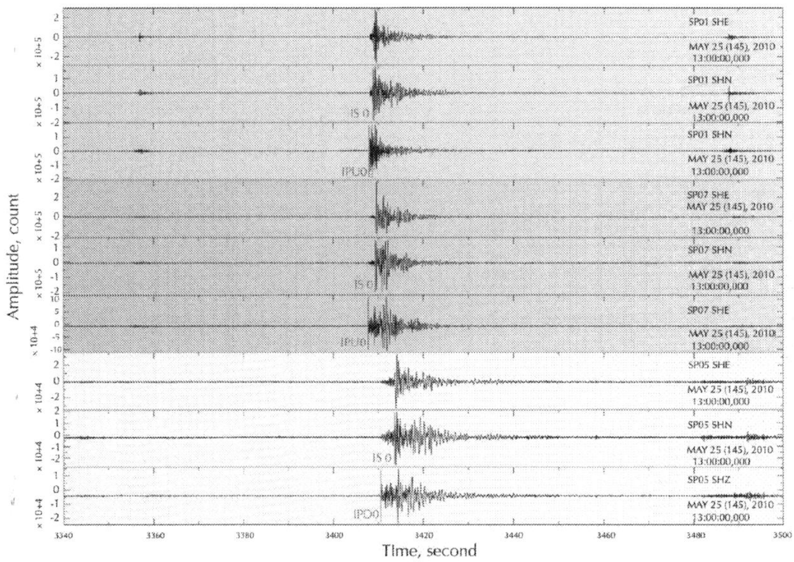

Figure 2: Example of seismic record of several aftershocks by 3 portable stations with indications of the P and S arrivals is presented.

METHODOLOGY

We used the Local Tomography Software (LOTOS), described in detail by Koulakov (2009) and freely available at http://www.ivan-art.com/science/LOTOS The version of the code used (version 12) allows for the simultaneous inversion of P-velocities (Vp) and S-velocities (Vs) (and Vp/Vs) and source coordinates.

The input data for the code were the arrival times of local seismicity and station coordinates. The calculations started with the absolute location of sources and estimates of an optimal 1D velocity model. In the version of the code used, the preliminary location of sources is based on a linear approximation of rays, which shortens the calculation procedure. The reference 1D model was defined using a constant Vp/Vs ratio and a fairly simple representation of the P-wave model. We used three to four velocity values at different depths, and velocity was linearly interpolated between the levels. The preliminary location step was performed for many different reference models. The model that produced the minimal root mean square (rms) of residuals and the maximal number of selected data was selected as a reference model for further 3D tomography modeling. In our final selected model, the Vp/Vs ratio is constant at 1.68 and P-wave velocities are defined at four levels (Vp =4.8 km/s at 0 km, 5.2 km/s at 10 km depth, 5.8 km/s at 20 km depth, and 7.0 km/s at 30 km depth).

The tomographic inversion was performed based on an iterative procedure. Each iteration started with source location in the updated 3D model using the bending algorithm for 3D ray tracing. The construction of the parameterization grid was performed only in the first iteration. The nodes of the grid were distributed in the volume of interest according to the ray density. The minimum grid spacing in the vertical and horizontal direction was 2 km. To reduce any effect of grid dependency upon the solution, we performed the inversions for four different grids having different basic azimuthal orientations. Note that the parameterizations used for the velocity representation and 3D ray tracing were completely independent. The inversion was performed simultaneously for P- and S-wave anomalies, source parameters (x, y, z, and t0), and station corrections. The solution was regularized using gradient damping between neighboring nodes. Inversion of the derived matrix was performed using the LSQR algorithm (Paige and Saunders 1982; Nolet 1987). Optimal values of damping and of weighting parameters were defined based on the results of synthetic modeling.

MODELING AND TOMOGRAPHY RESULTS

The analysis starts with estimating the optimal reference velocity model. We considered a number of different 1D velocity models with quite simple definitions (2–3 velocity values at different depth and constant Vp/Vs ratio) and selected among them a model with provides minimal average residuals and minimum of rejected data. In such model, the Vp/Vs ratio was equal to 1.69 and P-velocities at 0, 10, 20 and 30 km depth were 4.8, 5.2, 5.8 and 7.0 km/s, respectively. Between the depth levels, velocity distribution was linearly interpolated. For this velocity model we selected 1136 events with 9885 P-wave and 2165 S-wave picks, taking into account only those events having more than 8 recorded picks and with the criterion that the residuals computed in the 1D model should not exceed 0.8 s. The values of average residuals and variance reduction during three iteration steps are presented in Table 1. The first iteration corresponds to source location in the 1D starting model. After three iterations, the residuals reduce to 0.089 s and 0.067 s for P- and S-wave data, respectively, which is close to the value of picking accuracy. A significant reduction of residuals (41% and 55% for P- and S data), taking into account the relatively small distances between sources and receivers, indicates the high quality of data and the high variation in seismic properties.

Table 1: Values of standard deviation variance reduction for the P and S data after the step of source locations at different iterations

Iteration	P-residual deviation, s	P-residual reduction, %	S-residual deviation, s	S-residual reduction, %
1	0.153	0	0.150	0
2	0.090	40.7	0.070	53.6
3	0.089	41.4	0.067	55.4

Abacha et al.

Abacha et al. SpringerPlus 2014 3:650, doi:10.1186/2193-1801-3-650

To assess the resolution of the results, we performed a series of synthetic tests for which we simulated the conditions of the observed data inversion. In particular, we used identical source–receiver configurations as in the real data experiment. The synthetic travel times were computed using a 3D ray tracing algorithm based on the bending approach. After computing the synthetic data, we ignored the coordinates of the "true" sources, and performed a full calculation workflow, including the source locations, as in the real data case. Thus, in this modeling, we faced the problem of a trade-off between source and velocity parameters, both of which affect the results of passive-source tomography.Figure 3 presents the results of inversion for two checkerboard models composed of 3 × 3 (model "Board 1") and 2 × 2 (model "Board 2") patterns. The sizes of anomalies for these two models are 2 km and 3 km, and the spacings are 1 km and 2 km, respectively. The amplitudes of anomalies in all cases were ±5%. In these cases, the anomalies are unlimited with depth. The results of the checkerboard reconstruction show that in the focal area (Figure 3, center of the map), where most earthquakes are located, the boundaries between anomalies in the 2 × 2 case can be clearly discerned. The outer limits of the anomalies are strongly smeared due to the dominant orientations of rays. The test with the 3 × 3 model shows that the 2-km size anomaly can be resolved only in the center of the aftershock area, but the outer anomalies are strongly biased.To further investigate the effect of smearing, we performed 12 different tests with single synthetic anomalies of 3 × 3 km lateral size placed in different parts of the study region. In all these cases, the amplitude of anomalies was 5%, and the anomalies were unlimited with depth. Examples of reconstructions for six of these models are presented in Figure 4. For the case of locating the pattern in the aftershock area (model "Single_1"), the reconstruction reveals the limits of the anomaly correctly. However, for all other cases of placing the patterns outside the central area, the reconstructed anomalies are strongly smeared in the radial directions. This smearing effect should be taken into account when considering the results of data inversion.It is well known than the vertical resolution of passive tomography is usually poorer than the horizontal resolution, which can be explained by the effect of the trade-off between seismic velocity and source parameters. This effect is demonstrated in a series of tests presented in Figure 5. In this case, we defined several synthetic patterns across two vertical sections used for presenting the main results (Figure 6). All these eight plots represent different synthetic models, four for each of two sections; thickness of the anomaly in the direction across the section is 5 km. Here we present the reconstruction results for the P-anomalies. We analyzed different depth intervals of the anomalies, and the purpose of the test was to check whether the existing observation system is able to resolve any differences between these models. In fact, the reconstruction results shown in Figure 5 display a very strong vertical smearing of anomalies, which does not allow the upper and lower limits of the synthetic anomalies to be resolved. These tests show that care is needed when interpreting any vertical changes of structures in real data results.The results of tomographic inversion, which include P- and S-wave velocity anomalies, the Vp/Vs ratio, and the locations of sources, are presented in horizontal and vertical sections in Figures 6 and 7, respectively. We show the anomalies only in areas located at distances of less than 4 km from a nearest node of the parameterization grid which was set according to the ray density. Some anomalies in areas outside the network are related to the ray paths travelling to the remote stations of the Algerian network which were also involved in this study (Figure 8). Note that these outside patterns might be strongly smeared and thus should be considered with prudence. Both the P- and S-wave velocity structures appear to differ in shape from those in our model. The P-wave velocity structures are represented by contrasting patterns with radial orientations. However, as was shown by the synthetic tests, these shapes could be artifacts due to smearing outside the aftershock area. The S-wave velocity model is represented primarily by a single low-velocity anomaly in the central part of the study area that is surrounded by high-velocity anomalies. Interestingly, the amplitudes of the S-wave velocity anomalies are around 5%, which is lower than the value for the P-wave velocity anomalies (7%–9%). The S-wave velocity model shows a much larger reduction in variance and a smaller rms of remnant residuals after inversion than does the P-wave velocity model. Thus, the lower amplitude in the S-wave velocity model cannot be due to poorer quality of the S-wave data and consequent stronger damping in the inversion. On the basis of several inversion trials, we conclude that the obtained amplitudes are realistic. The Vp/Vs ratio model is obtained from the division of the resulting P- and S-wave velocities. This model displays a clear feature of high Vp/Vs ratio values, up to 1.78, in the area of the fault. In the surrounding areas, values are generally low (Vp/Vs =1.62–1.66).The resulting images of horizontal and vertical sections do not show significant variation in structure with depth. However, as was shown by the synthetic tests in Figure 5, the existing data do not allow a robust resolution of the vertical variation.

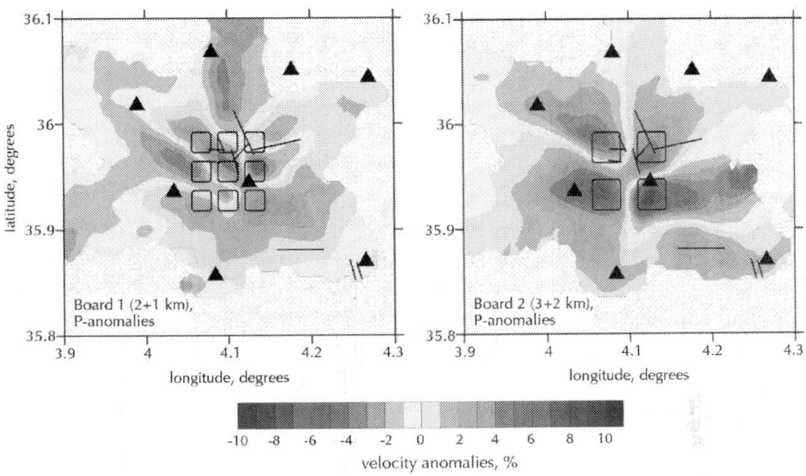

Figure 3: Inversion results for two checkerboard models composed of 3×3 and 2×2 patterns. The sizes of anomalies and the spacing is 2 + 1 km and 3 + 2 km, respectively. The reconstructions results are shown for the depth of 4 km. Black triangles indicate seismic stations and the black lines represent the major faults mentioned in the Figure 1b.

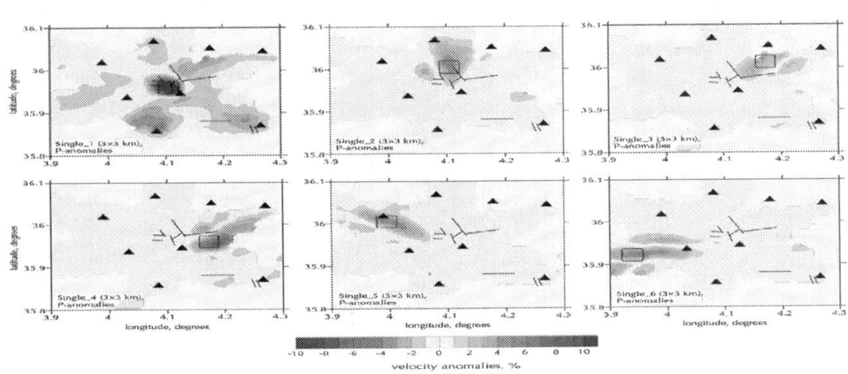

Figure 4: Examples reconstructions for six models with single synthetic anomalies 3×3 km of lateral size placed in different parts of the study region to further investigate the effect of smearing. In all these cases, the amplitudes of anomalies were 5%.

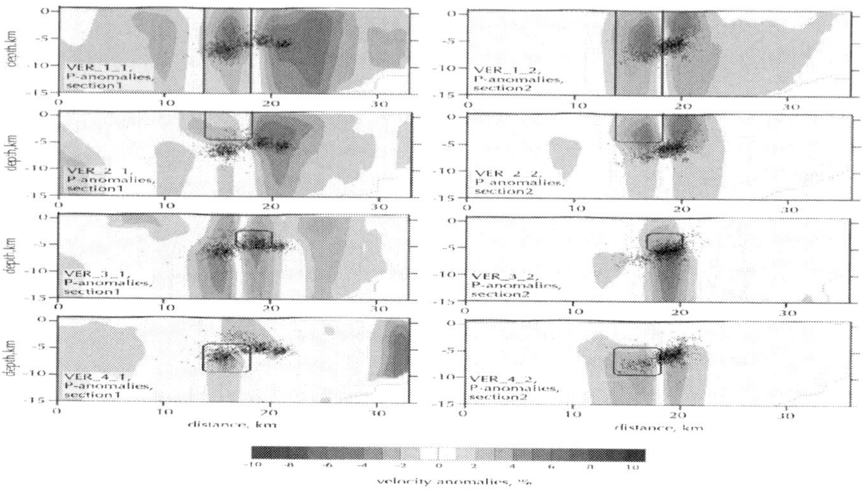

Figure 5: Examples of reconstructions of four different synthetic models defined along sections 1 and 2 with the position indicated in Figure6. Red stars are epicenters of the three main shocks.

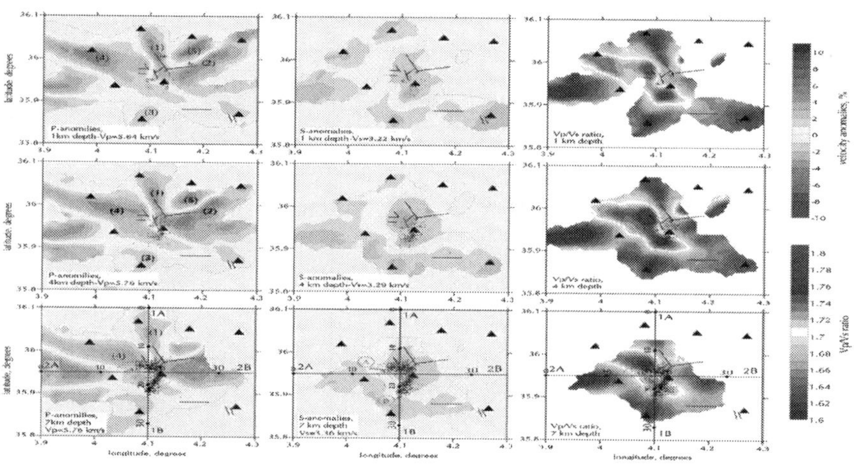

Figure 6: P, S-velocity anomalies and the Vp/Vs ratio in horizontal sections at 1, 4 and 7 km depth with the corresponding values of absolute reference velocities. Dots represent the locations of events. Black lines represent the major faults mentioned in the Figure 1b. Black triangles show seismic stations. Red stars are epicenters of the three main shocks. A and B are the two faults proposed according the aftershocks distribution.

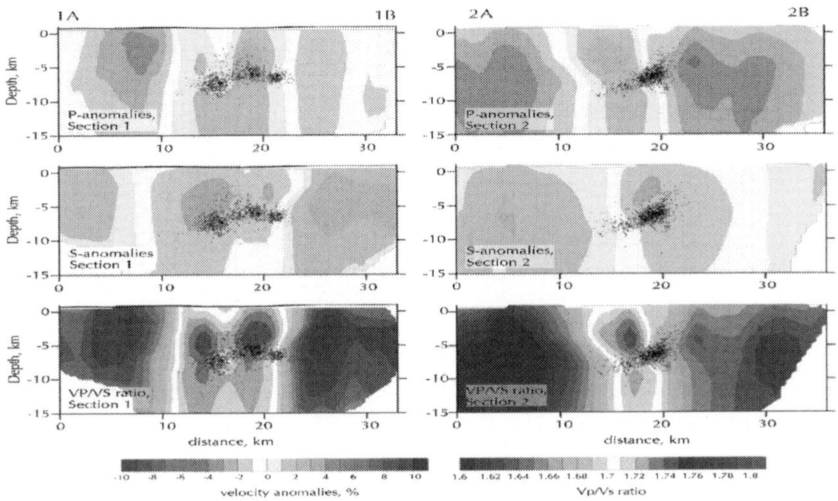

Figure 7: P, S-velocity anomalies and the Vp/Vs ratio in vertical sections 1 and 2 with the aftershocks relocated. Red stars are epicenters of the three main shocks.

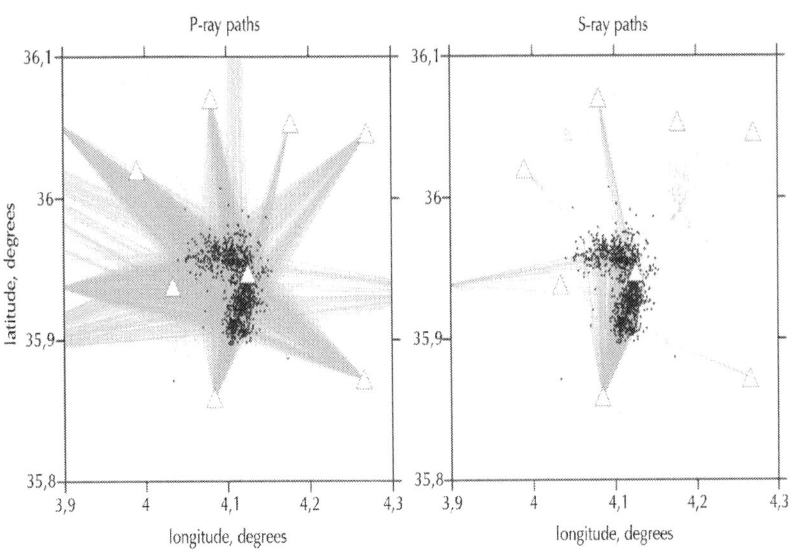

Figure 8: Path of the P and S rays for the source locations after three iterations of tomographic inversion. White triangles are the stations; black dots are the events.

DISCUSSION

First of all we should note that the accuracy of source locations is quite high due to the favorable deployment of seismic stations around the aftershock area. We observe a rather well-clustered swarm of seismicity that appears to be clearly aligned in two segments. In map view, the events form a "7"-shaped cluster, with one segment oriented E–W showing a fault segment about 7 km long and 3 km wide (A in Figure 6), and another oriented NNE–SSW showing a fault segment about 8 km long and 3 km wide (B in Figure 6).Two cross-sections were carried out show the vertical distribution of aftershocks and the variation in velocity with depth. Cross-section 1 (1A, 1B in Figure 6) is perpendicular to the first cluster, and cross-section 2 (2A, 2B in Figure 6) is nearly perpendicular to the second cluster. Figure 7 shows the P- and S-wave velocity anomalies and the Vp/Vs ratio in vertical sections 1 and 2 together with the relocated aftershocks. The first cluster (A in Figure 6) is oriented E–W and is distributed on a near-vertical plane 3 km wide (between 13 and 16 km), 8 km long (between 13 and 21 km, see cross-section 2), and between 5 and 9 km deep (see cross-section 1). The second cluster (B in Figure 6) is oriented NNE–SSW and is also distributed on a near-vertical plane 3 km wide (between 18 and 21 km), 8 km long (15 and 23 km, see cross-section 1), and between 3 and 7 km deep (see cross-section 2).

These results are in agreement to those of Yelles et al. (2013), who relocated the same data and found that the focal mechanisms of the second cluster (B) show near-vertical left-lateral strike-slip fault planes and that those of the first cluster (A) show a high-angle reverse fault.

The spatiotemporal evolution of the Beni-Ilmane aftershocks as defined by Yelles et al. (2013) confirms the three periods of rupture processes that were considered according to the occurrence of the three large main shocks of very similar magnitude (red stars 1, 2, and 3 in Figure 6). The first shock occurred on May 14 on a NNE–SSW-trending, near-vertical, left-lateral strike-slip fault, and may have provoked the second shock that occurred two days later on an E–W reverse fault. The third shock occurred south of the first one nine days after, with the same fault parameters.

No significant differences are observed between the velocity features at different depth levels. This is probably due to poor vertical resolution, as shown by the synthetic tests presented in Figure 5. The structure of P-wave velocity anomalies most likely represents shallow structures and can be associated with the surface geology (black lines in Figure 6). It is widely accepted that P-wave velocity is sensitive mainly to variations in the composition of rocks (e.g., Eberhart–Phillips et al. 1989; Behn and Keleman 2003). Therefore, the resulting P-wave velocity anomalies may represent geological structures of different origins. The low-velocity anomalies marked (1) and (2) in Figure 6 fit with the fault alignments determined from geological evidence (Baldini 1966). Pattern (1) may coincide with fault "a" and pattern (2) with fault "b" and the syncline of Beni-Ilmane (Figure 1b). These two patterns appear clearly for the first two horizontal sections at depths of 1 and 4 km, but start to disappear from the third section at 7 km depth. No correlation is found between these anomalies and the aftershock distributions that might be explained by a shallow expression of the faults corresponding to the occurrence of seismicity. However, it should be noted that care should be taken with these radial-shaped anomalies, because they can be caused partly by smearing, as was shown by synthetic tests (Figure 4).Anomaly (3) appears to be located at the prolongation of seismicity cluster B and may represent the end of the strike-slip fault where the second phase of aftershock activity occurred. P-wave high-velocity anomalies (4) and (5) may represent rigid blocks of upper crust that are not deformed by regional stresses. These anomalies are located at Jebel Dira and along of the northern part of Choukchot anticline, respectively (Figure 1b).

The S-wave velocity model appears to differ substantially from the P-wave velocity model. In contrast to the compressional velocity, the shear velocity is generally regarded as being related to the fracturing, porosity, and content of fluids (e.g., Domenico 1984; Eberhart-Phillips et al. 1989). We observe an S-wave low-velocity anomaly in the central part of the study region that coincides with the aftershock area (Figure 6). This pattern is especially clear in maps of the Vp/Vs ratio. Relatively high values of the Vp/Vs ratio (~1.78) compared with the surrounding areas (1.62–1.66) probably indicate the presence of high fluid contents. These fluids can be released from deeper levels due to fault movements and quickly migrate upwards. This hypothesis is supported by vertical sections (Figure 7) that show the major Vp/Vs anomalies located above the seismic clusters.

Oil exploration conducted in the South Tellian border region during the 1990s indicated that this area could contain important hydrocarbon reservoirs (Bracene 2002). Two hypotheses for the origin of the oil have been proposed. One considers the bedrock as Turonian (British Petroleum 1995: Petroleum evaluation of the Sour El Ghozlane, unpublished) and the other considers it to be upper Albian (Maxus 1994: Convention Sonatrach-Maxus sur L'Atlas Saharien central, unpublished). Our tomographic investigations through the negative S-wave velocity anomaly shown in Figure 6 seem to reveal oil and gas fields content in these levels.

CONCLUSIONS

The 2010 Beni-Ilmane earthquake sequence represents the most important recent seismic event in the transition seismotectonic zones of the Bibans-Hodna Mountains and the Tellian Chain High Plateaus, where the seismic activity is usually relatively low. This earthquake sequence was marked by three main shocks of very similar magnitude in 10 days, and by strong aftershock activity that lasted for several months. This active seismic phase was triggered by conjugate E–W reverse and N–S dextral faulting.

This study is the first tomographic investigation of the region. We performed inversion using the LOTOS code, which allows for the simultaneous determination of P- and S-wave velocity distributions as well as the relocation of aftershocks. Our selected model was thoroughly verified by various synthetic tests that show an acceptable horizontal resolution in the central part of the study region coinciding with the aftershock area. This is in contrast to the surrounding areas, where the patterns of anomalies are strongly smeared due to the dominant orientations of rays. We also demonstrated the limits of vertical resolution, which can be explained by the effect of the trade-off between seismic velocity and source parameters.

The P-wave velocity anomalies reveal low-velocity patterns that correspond to the fault alignments determined from geological evidence. The derived positive P-wave anomalies may represent rigid blocks of the upper crust that are not deformed by regional stresses.

The S-wave velocity model appears to differ considerably from the P-wave velocity model. We observed a low-velocity anomaly in the central part of the study region that coincides with the aftershock area. This pattern is especially clear in maps of the Vp/Vs ratio. Relatively high values of the Vp/Vs ratio (~1.78) in the central area compared with the surrounding areas (1.62–1.66) probably indicate the presence of high fluid contents. These fluids may be released from deeper levels by fault movements and migrate quickly upwards.

Reports of oil investigations conducted during the 1990s in the South Tellian border region show the existence of hydrocarbon reservoirs (both oil and gas). These reservoirs coincide with the low-velocity anomaly in the central part of the study region tomographic investigations. Then tomographic investigations seem to be a good tool to discuss and reveal presence of oil content in potential hydrocarbon region.

AUTHORS' CONTRIBUTIONS

AI collected the data, carried the processing, and participated in the written of this manuscript and the interpretation of the results. KI participated in the written, processing of data and the interpretation of the results. SF participated in the written and the correction of manuscript. YA participated in the written and the coordination between all authors. All authors are read and approved the final manuscript.

ACKNOWLEDGMENTS

The authors gratefully acknowledge the Editor Max Haring and the three anonymous reviewers for their useful comments and suggestions.

REFERENCES

1. Ayadi A, Dorbath C, Ousadou F, Maouche S, Chikh M, Bounif A, Meghraoui M (2008) Zemmouri earthquake rupture zone (*Mw* 6.8, Algeria): Aftershocks sequence relocation and 3D velocity model. J Geophys Res 113:B09301 doi:10.1029/2007JB005257
2. Baldini P (1966) Notice explicative de la carte géologique au 1/50 000 « TARMOUNT » feuille (140). Service géologique de l'Algérie, Alger.

3. Behn MD, Kelemen PB (2003) Relationship between seismic P-wave velocity and the composition of anhydrous igneous and meta-igneous rocks. Geochem Geophys Geosyst 4:1041 doi:10.1029/2002GC000393, 5
4. Beldjoudi H, Guemache MA, Kherroubi A, Semmane F, Yelles-Chaouche AK, Djellit H, Amrani A, Haned A (2006) The Laalam (Bejaia, Northeast Algeria) moderate earthquake (Mw:5.2). Pageoph. doi:10.1007/S00024-009-0462-9
5. Beldjoudi H, Delouis B, Heddar A, Nouar OB, Yelles-Chaouche AK (2011) The Tadjena earthquake (Mw =5.0) of December 16, 2006 in the Cheliff Region (Northern Algeria): waveform modelling, regional stresses, and relation with the Boukadir Fault. Pageoph 169:677-691 doi:10.1007/s00024-011-0337
6. Benouar D (1994) The Melouza earthquake of 21 February 1960, Seismicity of Algeria and adjacent region during the twentieth century. Ann. Geofis 37(4):862
7. Bounif A, Dorbath C (1998) Three dimensional velocity structure and relocated aftershocks for the 1985 Constantine, Algeria (M_s =5.9) earthquake. Annali di Geofisica 41(N.1):93-104
8. Bracene R (2002) Géodynamique du Nord de l'Algérie: Impact sur l'exploration Pétrolière. In: PhD. Thesis. Vol 1. France: Université. De Cergy Pontoise.
9. Chiarabba C, Amato A (1994) From tomographic images to fault heterogeneities. Annali di Geofisica XXXVII:N. 6
10. Chiarabba C, Amato A, Meghraoui M (1997) Tomographic images of the El Asnam fault zone, and the evolution of a seismogenic thrust-related fold. J Geophys Res 102:doi: 10.1029/97 JB01778. Issn: 0148–0227
11. Domenico S (1984) Rock lithology and porosity determination from shear and compressional wave velocity. Geophysics 49(8):1188-1195 doi:10.1190/1.1441748
12. Dorbath C, Oppenheimer D, Amelung F, King G (1996) Seismic tomography and deformation modeling of the junction of the San Andreas and Calaveras faults". J Geophysical Res: Solid Earth (1978–2012) 101(Issue B12):27917-27941 doi:10.1029/96JB02092

13. Eberhart–Phillips D, Han D, Zoback M (1989) Empirical relationships among seismic velocity, effective pressure, porosity, and clay content in sandstone. Geophysics 54(1):82-89 doi:10.1190/1.1442580

14. Kieken M (1974) Etude géologique du Hodna, du Titteri et de la partie occidentale des Biban.

15. Kieken M (1975) Etude géologique du Hodna, du Titteri et de la partie occidentale des Biban.

16. Koulakov I (2009) LOTOS code for local earthquake tomographic inversion. Benchmarks for testing tomographic algorithms. Bull Seism Soc Am 99(1):194-214 doi:10.1785/012008001

17. Koulakov I, Bindi D, Parolai S, Grosser H, Milkereit C (2010) Distribution of seismic velocities and attenuation in the crust beneath the North Anatolian Fault (Turkey) from local earthquake tomography. Bull Seism Soc Am 100(1):207-224

18. Nolet G (1987) Seismic wave propagation and seismic tomography. In: Nolet G (ed) Seismic Tomography, Dordrecht: Reidel. pp 1-23

19. Paige CC, Saunders MA (1982) LSQR: an algorithm for sparse linear equations and sparse least squares, ACM transactions on mathematical software. Trans. Math. Softw 8:43-71

20. Salah MK, Sahin S, Destici C (2007) Seismic velocity and Poisson's ratio tomography of the crust beneath southwest Anatolia: an insight into the occurrence of large earthquakes. J Seism 11:415-432 doi:10.1007/s10950-007-9062-2

21. Salah MK, Sakir Sahin S, Aydin U (2011) Seismic velocity and Poisson's ratio tomography of the crust beneath East Anatolia. J Asian Earth Sci 40:746-761 doi:10.1016/j.jseaes.2010.10.021

22. Thurber C, Roecker S, Ellsworth W, Chen Y, Lutter W, Sessions R (1997) Two-dimensional seismic image of the San Andreas Fault in the Northern Gabilan Range, central California: evidence for fluids in the fault zone". Geophys Res Lett. 1591-1594 doi:10.1029/97GL01435

23. Yelles K, Lammali K, Mahsas A, Calais E, Briole P (2004) Coseismic deformation of the May 21st, 2003, Mw =6.8 Boumerdes earthquake, Algeria, from GPS measurements. Geophys Res Lett 31:L13610doi:10.1029/2004GL019884

24. Yelles K, Abacha I, Semmane F, Beldjoudi H (2013) The Beni-Ilmane (North-Central Algeria) Earthquake Sequence of May 2010. Pageoph.in press

25. Yolsal-Çevikbilen S, Biryol CB, Beck S, Zandt G, Taymaz T, Adıyaman HE, AA O (2012) 3-D crustal structure along the North Anatolian Fault Zone in north-central Anatolia revealed by local earthquake tomography. Geophys J Int 188:819-849doi:10.1111/j.1365-246X.2011.05313.x

Citations

CHAPTER 1

Cunbao Li, Lingzhi Xie, Li Ren, Heping Xie, and Jun Wang, "Evaluating the Applicability of Fracture Criteria to Predict the Crack Evolution Path of Dolomite Based on SCB Experiments and FEM," Mathematical Problems in Engineering, vol. 2013, Article ID 959806, 13 pages, 2013. doi:10.1155/2013/959806.

CHAPTER 2

Guoping Jiang, Wei Shi, and Lili Huang, "Fractal Analysis of Permeability of Unsaturated Fractured Rocks,"The Scientific World Journal, vol. 2013, Article ID 490320, 5 pages, 2013. doi:10.1155/2013/490320.

CHAPTER 3

Tielin Chen, Liangyi Zhang, and Dingli Zhang, "Simulation of Hydraulic Fracture in Unsaturated Soils with High Degree of Saturation," Advances in Materials Science and Engineering, vol. 2014, Article ID 981056, 10 pages, 2014, doi:10.1155/2014/981056.

CHAPTER 4

Arne Stormo, Olivier Lengliné, and Jean Schmittbuhl, Mechanical Origin of B-value Changes during Stimulation of Deep Geothermal Reservoirs, doi:10.1186/s40517-014-0022-0.

CHAPTER 5

Thea K Hincks, Jean-Christophe Komorowski, Stephen R Sparks and Willy P Aspinall, Retrospective Analysis of Uncertain Eruption Precursors at La Soufrière Volcano, Guadeloupe, 1975–77: Volcanic Hazard Assessment using a Bayesian Belief Network Approach, doi:10.1186/2191-5040-3-3.

CHAPTER 6

Paul A Siratovich, Michael J Heap, Marlène C Villenueve, James W Cole, and Thierry Reuschlé, Physical Property Relationships of The Rotokawa Andesite, a Significant Geothermal Reservoir Rock in the Taupo Volcanic Zone, New Zealand, doi:10.1186/s40517-014-0010-4.

CHAPTER 7

Issam Abacha, Ivan Koulakov, Fethi Semmane ,and Abd Karim Yelles-Chaouche, Seismic Tomography of the Area of the 2010 Beni-Ilmane Earthquake Sequence, North-Central Algeria, doi:10.1186/2193-1801-3-650.

Index